Introduction to Data Science for Engineering Students

Introduction to Data Science for Engineering Students

ILIAS BILIONIS

Purdue University, USA

W🌐 World Scientific

NEW JERSEY · LONDON · SINGAPORE · BEIJING · SHANGHAI · HONG KONG · TAIPEI · CHENNAI · TOKYO

Published by

World Scientific Publishing Co. Pte. Ltd.

5 Toh Tuck Link, Singapore 596224

USA office: 27 Warren Street, Suite 401-402, Hackensack, NJ 07601

UK office: 57 Shelton Street, Covent Garden, London WC2H 9HE

Library of Congress Cataloging-in-Publication Data

Names: Bilionis, Ilias author

Title: Introduction to data science for engineering students / Ilias Bilionis, Purdue University, USA.

Description: New Jersey : World Scientific, [2026] | Includes bibliographical references and index.

Identifiers: LCCN 2025035041 | ISBN 9789819822430 hardcover |
  ISBN 9789819823420 paperback | ISBN 9789819822447 ebook |
  ISBN 9789819822454 ebook other

Subjects: LCSH: Engineering--Data processing

Classification: LCC TA345 .B528 2026 | DDC 620.001/51--dc23/eng/20251219

LC record available at https://lccn.loc.gov/2025035041

British Library Cataloguing-in-Publication Data

A catalogue record for this book is available from the British Library.

For any available supplementary material, please visit
https://www.worldscientific.com/worldscibooks/10.1142/14563#t=suppl

Desk Editors: Nambirajan Karuppiah/Veronica Lee

Typeset by Stallion Press
Email: enquiries@stallionpress.com

To Katerina, Philippos, and Marietta

Preface

I wrote this book to provide an introduction to data science for engineering students without prior knowledge. The material included here is part of a one-credit sophomore course I have developed and taught for the Purdue University School of Mechanical Engineering. The course includes only one lecture per week. One chapter of this book corresponds, more or less, to one lecture.

The book will not teach you everything about data science. But, it will get you through the basics so you can get your hands dirty with real data. I start by defining data science and demonstrating why engineers should care. Most data science nowadays happens in Python. So, I cover the basics of Python, from simple mathematical expressions to data loading and visualization. Data science has to deal with uncertainty and randomness, so I introduce several concepts from probability theory. I use these probabilistic concepts to summarize and compare datasets.

Data science is all about making models. I show you how to make some simple models. For example, using experimental data, I demonstrate how to create a model that predicts engine efficiency based on operating conditions. I also teach you how to evaluate the models and talk a little about how you should think when you use the model predictions to make important decisions.

After completing this book, you should be able to:

- Program in Python within a Jupyter Notebook environment.
- Summarize and compare datasets using empirically estimated statistics.

- Summarize and compare datasets visually.
- Represent uncertainty using probabilities.
- Apply simple probability rules to propagate uncertainty.
- Estimate probabilities from data.
- Solve regression problems (learn from data a linear model that takes you from a set of input variables to a continuous variable).
- Solve classification problems (learn from data a model that takes you from a set of input variables to a discrete label).

You will have no trouble following the book if you understand calculus and matrix-vector multiplication. Some prior programming experience will also be helpful.

Now, this book reflects my personal preferences, biases, and style. Lots of people will likely disagree with the choices I made. To the extent that I understand my own biases, they are as follows.

- I like mathematical proofs and derivations, but not at the same rigor level as in a math textbook. I like what you would call "physics-style" proofs.
- I want my work to be replicable. So, I provide the code replicating all the results and figures for this book. And I mean *all* of it.
- I am religiously Bayesian. That is, I like to use probabilities to represent uncertainty and train models using Bayesian inference. You won't see much of that here, but it explains why I insist on some derivations.

As you work through the book, you will notice that some sections are marked with a Colab tag. If you have an electronic version of this book, you can click on the tag, and it will take you to a Jupyter Notebook that you can run on your browser. The notebooks are hosted on Google Colab — so you need a Google account to run them. These notebooks replicate all the content of the corresponding section. They allow you to run the code yourself and typically include a few questions requiring you to modify the code and run it again. This is the best way to learn the material.

If you don't have an electronic version of this book, you can still run the code in the notebooks. You have two options. The first way is to access them through the free online version of the book accessible at https://purduemechanicalengineering.github.io/me-239-intro-to-data-science/.

This online version is less polished, but the Jupyter notebooks are the same. You will have to find the notebook you are interested in manually. Once you find it, you can launch it on Colab by clicking at the rocket icon in the top right corner of the notebook. You will need to have a Google account to run the notebooks.

The second way is to download the notebooks from the GitHub repository at https://github.com/PurdueMechanicalEngineering/ me-239-intro-to-data-science.

You can then run them locally. This is a bit more cumbersome, but it is a good way to learn how to use Jupyter notebooks, and you will have complete control over the environment.

Finally, throughout the book, we will use data from the GitHub repository. When running the notebooks on Colab, you won't have to worry about the data as I download them. When running the notebooks locally, you must download the data yourself. You can find the data in the `data` folder of the GitHub repository.

I hope you enjoy the book and get motivated to learn more about data science!

I. Bilionis
Athens, Greece
June 2025

About the Author

Ilias Bilionis is a Professor of Mechanical Engineering at Purdue University, where he leads the Predictive Science Laboratory focused on AI technologies for accelerating engineering innovation. His interdisciplinary research has been funded by NSF, NASA, DARPA, AFOSR, and leading industry partners, including Ford, Cummins, and Eli Lilly. He created Purdue's data science curriculum for mechanical engineers, including the course on which this book is based. He is a recipient of the Outstanding Faculty Mentor Award, the Outstanding Engineering Teacher Recognition, and the Online Education Award from Purdue University.

Acknowledgments

First and foremost, I would like to thank all the people who support the Python community and make open-source software possible. In particular, the team behind the Project Jupyter and the Executable Books Project. I would also like to thank the Google Colab team for making it possible to run Jupyter notebooks in the cloud free of charge. The interactive style of this book would not have been possible without them!

I thank the students who have taken my ME 239 class at Purdue University. Their feedback has been invaluable in shaping the content of this book. In particular, I would like to thank Max Bolt for being the first undergraduate to ever work through all the material. Also, special thanks to Jitesh Panchal for encouraging me to develop the ME 239 course in the first place and Eckhard Groll for being supportive during my first experiments with the material.

I would like to thank NSF grant 2347472 and Purdue University for supporting the development of this book.

Lastly, I thank my brother, Stavros Bilionis, and his team for creating the cover art.

Contents

15. Linear Regression   289

Chapter 1

Introduction to Data Science

1.1 Learning Objectives

In this chapter, I define data science and highlight some of its applications in engineering. I explain the lecture book structure and discuss why Python is my programming language of choice. I introduce Jupyter notebooks and demonstrate how to run them on Google Colab. The chapter covers Python fundamentals including expressions, names, basic function calling, and data types. Finally, I show how to get help documentation for Python functions.

1.2 What Is Data Science?

According to Wikipedia[1]

> Data science is an inter-disciplinary field that uses scientific methods, processes, algorithms and systems to extract knowledge and insights from [...] data.

This is a pretty good definition!

In engineering applications, "scientific methods" come in the form of physical laws. So, in our context, data science is about combining domain knowledge (dynamics, mechanics of materials, fluid mechanics, etc.) with data. This connection between data and domain knowledge is not ad hoc. The "glue" is provided by probability theory and

---

[1]https://en.wikipedia.org/wiki/Data_science.

1

statistics. This lecture book is giving you the foundations you need to start learning more, or more precisely, the foundation we can fit in 16 lectures.

1.3 Why Should You Care?

What follows is an incomplete list of cool things that one can do combining engineering with data science. Note that these applications are rather advanced. We will not learn how to carry them out in this book. You will, however, learn the fundamentals upon which the data science components of these applications are based.

1.3.1 *SpaceX Falcon rocket landing*

We have all watched in awe how SpaceX rockets can land autonomously on a platform that catches them (Figure 1.1).[2] This advanced feat is possible thanks to the use of data science techniques combined with physical models and control theory. The first step is that a data science technique called *filtering*, which is used to estimate the state of the rocket (position, velocity, etc.) using noisy data coming from GPS, accelerometers, cameras, and other sensors. The technique combines the data with physical models that describe the rocket dynamics (essentially Newton's laws of motion). It finds the best estimate of the rocket state that is consistent with the data and the physical models. Think for a while what you will need to do to pull this off in real-time! You need reliable sensors that can withstand the harsh conditions of the rocket launch and landing. You need to model the rocket dynamics accurately. You need communication and data acquisition systems to collect the data. And you need computational resources to run the filtering algorithm.

    After they have estimated the state, they can use it to decide which thrusters to activate to make the rocket follow an optimal trajectory. This is the control problem. The solution to this problem must be robust to the uncertainties in the rocket state and

---

[2]https://youtu.be/RYUr-5PYA7s.

Figure 1.1. SpaceX Starship ignition during its launch on IFT-5. By Steve Jurvetson — Flickr, CC BY 2.0, https://commons.wikimedia.org/w/index.php?curid=153992086.

the dynamics. The data science output is the input to the control problem.

1.3.2 *Boston dynamics spot autonomous navigation*

Spot is a dog robot (Figure 1.2).[3] Spot uses camera data to figure out where it is and navigate autonomously. Again, here we have a filtering problem and a control problem.

---

[3]Boston dynamics spot: https://youtu.be/4s7_ng-pSow.

Figure 1.2. Early version of the Big Dog robot. By DARPA — This file was derived from: DARPA Strategic Plan (2007).pdf, Public Domain, https://comm ons.wikimedia.org/w/index.php?curid=20798332.

1.3.3 *Smart extraterrestrial habitats*

The Resilient Extraterrestrial Habitats Institute (RETHi)[4] was a NASA-funded project based at Purdue. The vision of the institute was to develop technologies that enable the design of habitats on the Moon and Mars. These structures will likely remain without crew for significant periods of time ranging from months to years making autonomy a key requirement.

I was leading the Awareness Thrust of the institute. Here, as in the previous two examples, we need to estimate the state of the

---

[4]NASA-Purdue RETHi: https://youtu.be/yFd8wE9qtkw.

system (the health state of the habitat) using data from sensors. We also need to control the robotic agents to carry out maintenance and repair activities.

1.4 How to Use This Book?

This book contains math and code. The math is there to give you a solid foundation in the theory of data science. The code is there to give you a practical way to apply the theory. Read the math, with a pen and paper. Do the math exercises.

The code is another story. You can, of course, read the code. But that is not enough. You need to run the code. Along with the book, we provide Jupyter notebooks that contain the code in a format that you can run easily. Jupyter notebooks are a powerful tool that allows you to combine code, math, and text in a single document. Each section of the book contains a link to the corresponding Jupyter notebook. You can either download the notebooks and run them on your own computer (e.g., using an IDE like VSCode or Cursor), or run them on Google Colab, a free cloud service.

The Jupyter notebooks will contain questions. To answer them, you will need to modify the code and run it again. Do them all!

1.5 Why Python?

Throughout this book, we will use Python as our programming language of choice. Python is the programming language used in the majority of data science applications. Google is using Python (see https://docs.jax.dev). Meta is using Python (see https://pytorch.org). We use Python because it is absolutely free, very easy to learn, and has amazing libraries for pretty much anything you may need to do related to data science. Most importantly, it is fun!

1.6 Python Expressions: [Colab]

This is a good opportunity to introduce the interactive content of this book. Note the [Colab] link in the name of the section. If you use the electronic version of this book, you can click on the link and it will open the corresponding Jupyter notebook in Google Colab. Alternatively, you follow along by typing the commands in a Python interpreter[5] or in your own Jupyter notebook.

1.6.1   *Arithmetic operators*

The starting point is to use Python as a calculator. Here are some examples:

```
>>> 1 + 2
3
>>> 1 + 2 + 3
6
>>> 1 + (2 + 3)
6
>>> 2 * 3
6
>>> 2 * (3 + 4)
14
```

That was easy... The only thing to remember here is that whatever you enclose in parentheses is evaluated first. Just like regular mathematics.

The addition + and the multiplication * are called *binary operators* because they take two *operands*, i.e., two inputs. Let's continue with some more binary operators:

```
>>> 2 ** 3
8
>>> 5 ** 2
25
>>> (1 + 2) ** 2
9
```

So, the binary operator ** exponentiates. Remember this. It is different from other programming languages.

---

[5]You can start the Python interpreter by running the command **python** in your terminal if you're using a Mac or Linux machine. If you are using a Windows machine, you can use the command **python** in the Anaconda Prompt.

Here is the division operator:

```
>>> 2 / 3
0.6666666666666666
>>> 1 / 2
0.5
>>> 5 / (2 + 3) ** 2
0.2
```

What is the order in the previous expression? First, addition, then exponentiation, then division.

Okay, let's now divide with zero:

```
>>> 1 / 0
Traceback (most recent call last):
  File "<stdin>", line 1, in <module>
ZeroDivisionError: division by zero
```

Oops! This is how error messages look in Python. You need to get used to reading them. Look what it says: `ZeroDivisionError`. Well, it's obvious what it means. Read the errors! They look ugly, but they are there to help you!

The binary operator / is the division operator. What if we wanted integer division? Then you need to use the // operator:

```
>>> 3 // 2
1
>>> 9 // 2
4
>>> 6 // 4
1
```

Integer division by zero?

```
>>> 1 // 0
...
ZeroDivisionError: integer division or modulo by zero
```

We expected that!

What about the remainder of the division? This is the so-called modulo operator represented by the \% symbol. For example, if we divide 3 by 2, the remainder is 1. Here are some examples:

```
>>> 3 % 2
1
>>> 9 % 2
1
>>> 6 % 4
2
>>> 6 % 2
0
```

Try to divide by zero again. Can you guess the error message?
What about negative numbers? Sure,

```
>>> -4 + 5
1
>>> -5 + 5
0
```

1.6.2   *Writing numbers*

You can use scientific notation to define numbers in Python. For
example, the following are all perfectly fine Python expressions:

```
>>> 1e-1
0.1
>>> 2.5e-3
0.0025
```

Another thing that is useful to know is that you can use underscores to make numbers more readable. For example, the following
are all perfectly fine Python expressions:

```
>>> 1_000_000
1000000
```

1.6.3   *Rounding numbers*

Very often, we want to round numbers. We can use the built-in
`round()` function to do this. Here are some examples:

```
>>> round(2.24345)
2
```

```
>>> round(2.233535, 3)
2.234
```

How does `round()` work? You can use the help to figure it out:

```
>>> help(round)
Help on built-in function round in module builtins:

round(number, ndigits=None)
    Round a number to a given precision in decimal digits.

    The return value is an integer if ndigits is omitted or
    None. Otherwise, the return value has the same type as the
    number. ndigits may be negative.
```

`round()` is our first example of a Python function. As a matter of fact, it is a *built-in* Python function. We will learn quite a few of them as we go. By the way, `help()` is another built-in Python function. When in doubt about how a function works, use `help()` on it!

1.6.4 *Standard mathematical functions*

Python has some built-in math functions. They are organized in a python *module* called math. You can think of a Python module as a bag of related functions and variables. To use a module, you need to *import* it. Here is how you can import the functionality of the math module. You just do:

```
import math
```

Now you can use the contents of the math module. Here are some examples:

```
>>> math.pi
3.141592653589793
>>> math.sin(0.0)
0.0
>>> math.sin(math.pi / 2)
1.0
>>> math.cos(math.pi / 3)
0.5000000000000001
```

```
>>> math.cos(math.pi / 3) ** 2 + math.sin(math.pi / 3) ** 2
1.0
>>> math.sqrt(2)
1.4142135623730951
>>> math.sqrt(2) ** 2
2.0000000000000004
```

Note that some of the results are not coming out exactly right. There are tiny errors. These kinds of errors are called *floating point errors* or *numerical errors*. They arise because computers only store numbers with a finite precision. They are not a problem usually, but in data science, they can be a problem — especially when you add many small numbers (or numbers with different magnitudes together). Get used to them...

We saw how we can get π. How about e, the base of the natural logarithm?

```
>>> math.e
2.718281828459045
```

What about e^x? You should use the `math.exp` function in this case:

```
>>> math.exp(2.0)
7.38905609893065
```

What about e^{-2}?

```
>>> math.exp(-2.0)
0.1353352832366127
```

What about $1/e^{-2}$?

```
>>> 1 / math.exp(-2.0)
7.408182206817178
```

What happens to e^x when x is too big?

```
>>> math.exp(10)
22026.465794806718
>>> math.exp(100)
2.6881171418161356e+43
>>> math.exp(1000)
Traceback (most recent call last):
```

```
  File "<stdin>", line 1, in <module>
OverflowError: math range error
```

What on earth is an `OverflowError`? It is an error that arises when a number is too big to be represented in the floating point format. Python is using a set of bits to represent numbers. There is a limit to how big a number can be. Here, we are trying to get a number that is higher than that limit.

We are not going to cover all the functions in the math module. If you want to learn more about it, visit the math module documentation. Alternatively, you could use the help command `help(math)` to get the same information. If you are using a Jupyter notebook, you can just type `math.` and press `Tab` to see the list of functions.

If you do `help(math)`, notice that at the very end, you see:

```
...
DATA
e = 2.718281828459045
inf = inf
nan = nan
pi = 3.141592653589793
tau = 6.283185307179586
```

Okay, we know `pi` and `e`. The variable `tau` is just 2π. What about the other two? You need to know about them because you may encounter them in calculations. Let's start with `inf`, which represents infinity. It has the expected properties:

```
>>> math.inf
inf
>>> 1 / math.inf
0.0
>>> math.exp(math.inf)
inf
>>> math.exp(-math.inf)
0.0
>>> math.inf + math.inf
inf
>>> - math.inf
-inf
```

What about `nan`? This is another special number. It stands for "not a number." How can you get a result of an arithmetic operation that is not a number? Hmmm, how about this:

```
>>> math.inf - math.inf
nan
```

Or this:

```
>>> math.inf / math.inf
nan
```

You don't want to see `nan` in your calculations. If you do, there is something wrong and you need to stop right there and start fixing it!

1.7 Python Variables and Types: Colab

1.7.1 *Numerical data types*

We have already played with two kinds `int` and `float`. `int` stands for integer and `float` stands for *floating point number*, an approximation to a real number. You can use the built-in function `type()` to see the type of something. Here are some `int`'s:

```
>>> type(1)
<class 'int'>
>>> type(2)
<class 'int'>
```

And here are some `float`'s:

```
>>> type(1.24235)
<class 'float'>
>>> type(1e-5)
<class 'float'>
>>> type(1.0)
<class 'float'>
>>> type(2.)
<class 'float'>
```

When you add or multiply two `int`'s, you get an `int`:

```
>>> type(1 + 2)
<class 'int'>
>>> type(2 * 4)
<class 'int'>
```

Same thing for two `float`'s:

```
>>> type(1.0 + 2.3)
<class 'float'>
>>> type(2.1 * 3.5)
<class 'float'>
```

But what happens when you mix `int`'s and `float`'s? Let's try it out:

```
>>> type(1 + 2.4)
<class 'float'>
```

Adding an `int` to `float` gave you a `float`. Let's try another example:

```
>>> type(2 * 3.4)
<class 'float'>
```

Multiplying an `int` with a `float` gave you a `float`. This is a general rule: "When you have an operation that involves both `int`'s and `float`'s, the `int`'s are *promoted* to `float`'s and then the operation is carried out. Thus, the result is always a `float`."

What do we mean by promoted? We mean that when Python sees: `1 + 2.4` it changes it first to `2.0 + 2.4` and then carries out the addition. Python can do this by using the `float()` function. Here is an example:

```
>>> float(1)
1.0
>>> type(float(1))
<class 'float'>
```

Type promotion happens automatically in Python whenever you are trying to mix a more general type with a more restricted type.

There is also type *demotion*. For example, you can demote a `float` to an `int`:

```
>>> int(23.4353)
23
```

```
>>> type(int(23.33453))
<class 'int'>
```

Let's see what type is `math.inf` and `math.nan`:

```
>>> import math
>>> type(math.inf)
<class 'float'>
>>> type(math.nan)
<class 'float'>
```

Interesting! `math.inf` and `math.nan` are `float`'s!

1.7.2   *Python variables*

Variables are used to store values so that you can use them later in your code. Like this:

```
>>> x = 1
>>> y = 2
>>> z = x + y ** 3
>>> z
9
```

Variables in Python do not have types. It is the values stored in the variables that has a types. For example, the variable x has the value 1, which is an `int`. So, if you try `type(x)` you get:

```
>>> type(x)
<class 'int'>
```

But this type is not bound to the variable x. You can change it by reassigning x to a different value:

```
>>> x = 1.0
>>> type(x)
<class 'float'>
```

Now x is a `float`. This is very different from traditional programming languages where variables have types and you cannot change them. It is a very powerful feature of Python, because it allows you to write more flexible code. But it is also a feature that can lead to bugs if you are not careful.

1.7.3 *Boolean types*

When something is either true or false, we call it a *Boolean*. The corresponding Python type is `bool`. In Python, Boolean types are either `True` or `False`. As a matter of fact, `True` and `False` are special Python keywords. Here it is

```
>>> True
True
>>> False
False
>>> type(True)
<class 'bool'>
>>> type(False)
<class 'bool'>
```

What can you do with `True` and `False`? We will learn more about them when we talk about conditionals, but here are some examples:

```
>>> True and True
True
```

So `and` is another Python keyword...

```
>>> True and False
False
>>> True or False
True
>>> False or False
False
```

So `or` is another Python keyword... Here is the final example:

```
>>> not True
False
>>> not False
True
```

So `not` is another Python keyword...
You can also have longer expressions with `bool`'s:

```
>>> True and (False or True)
True
```

The order of operations is: parenthesis first, then `not`, then `and`, then `or`.

Just like for numerical types, there is a `bool()` function that can turn something into a `bool` type. For example:

```
>>> bool(1)
True
>>> bool(0)
False
>>> bool(1.0)
True
>>> bool(0.2242)
True
```

1.7.4 The `NoneType` and `None`

Python has a special value called `None` used to represent "nothing." Here it is

```
>>> None
```

Note that when you try to see `None` in the Python shell or in a Jupyter notebook cell, you do not see anything. This is because it is representing "nothing." You cannot see "nothing." What is the type of `None`?

```
>>> type(None)
<class 'NoneType'>
```

Exercises

1. Write Python code that evaluates the magnitude of the vector:

$$\vec{r} = 4\vec{i} + 3.5\vec{j} + 2.5\vec{k}.$$

2. Try to guess the output of the following expressions, then check your answers in Python:

```
>>> 1 / 2.0
>>> 2 // 3
>>> 2.0 // 3
```

```
>>> 5 % 3
>>> 5.0 % 3
>>> int(5.0 % 3)
```

3. Write Python code to evaluate the angle between the two vectors:

$$\vec{r}_1 = 4\hat{i} + 3.5\hat{j} + 2.5\hat{k},$$

and

$$\vec{r}_2 = 1.5\hat{i} + 2.5\hat{j}.$$

Remember that the angle between two vectors is:

$$\theta = \cos^{-1}\left(\frac{\vec{r}_1 \cdot \vec{r}_2}{|\vec{r}_1||\vec{r}_2|}\right).$$

Hint: Define variables x1,x2,y1,y2,z1,z2 and use math.acos().

4. Use Python to evaluate the following expression:

$$\frac{\sin(\pi/3)e^{-2}}{23^2}.$$

5. Find the type of the following Python variables. *Hint*: Copy-paste the code and use type() to find the types. You don't have to guess them. Identify some types we haven't seen yet. Look them up in the documentation.

```
# a
x = 4
# b
y = 23.5
# c
z = 4 / 3.1
# d
a = (1, 2, 4)
# e
b = [3, 4, 6]
# f
c = x > 2
# g
d = 'Hello there!'
```

```
# h
e = {'Name': 'Ilias Bilionis',
     'Phone number': 2106012120,
     'Zip code': 47906}
# i
import numpy as np
f = np.zeros((10, 4))
```

Chapter 2

Data Arrays

In this chapter, I introduce the sequence data types, tuples, lists, and numerical arrays. I demonstrate how I can use basic indexing to access elements of these sequences. I also introduce numerical arrays and demonstrate basic math with them.

2.1 Python Tuples: Colab

A sequence is a bunch of ordered objects. Ordered in the sense that you can say, "this is first," and "this is second." Obviously, sequences are very useful for data science. Data are made out of sequences of stuff.

There are two ways you can have sequences in Python. You can either have `tuple`'s or `list`'s. The difference between the two is that `tuple` are frozen (you cannot change them once you make them) but you can add or remove items for `list`'s. We look at tuples first.

Right now, you only know of `int`'s, `float`'s, `bool`'s and `NoneType`. So, that's all I will use to make tuples. But you can put anything in a tuple. Here is a tuple with two integers:

```
>>> x = (1, 2)
```

A tuple with two floats:

```
>>> y = (2.1232, 3.14)
```

You can mix and match types. Here is a tuple with two integers and one float:

```
>>> z = (5, 0, 34.34)
```

Type them in and use the `type()` function to see the types of `x`, `y`, and `z`. They are all `tuple`[1]:

```
>>> type(x), type(y), type(z)
(<class 'tuple'>, <class 'tuple'>, <class 'tuple'>)
```

You can ask a tuple how many elements it has using the function `len`:

```
>>> len(x), len(y), len(z)
(2, 2, 3)
```

2.1.1 *Indexing tuples*

You can access any element of a tuple using indexing. Note that indices start at 0 and go up to the number of elements in the tuple minus one. Let's try it on the `z` tuple defined above:

```
>>> z[0]
5
>>> z[1]
0
>>> z[2]
34.34
```

Let's be a bit more adventurous and try to access an element that is out of bounds:

```
>>> z[3]
Traceback (most recent call last):
  File "<stdin>", line 1, in <module>
IndexError: tuple index out of range
```

You get an `IndexError`. This is a very common bug. Remember it!

---

[1]What do we have there? What is this `something()`, `something()`, `something()`? This is called a *tuple packing*. You can separate Python expressions with commas and the results become a tuple. So, `type(x)`, `type(y)`, `type(z)` gave us actually a tuple of `type` objects!

Let's continue being adventurous. Try negative indices:

```
>>> z[-1]
34.34
>>> z[-2]
0
>>> z[-3]
5
```

That worked! Negative indices give you the elements starting from the last one. In particular, z[-1] gives you the last element of the tuple, z[-2] gives you the second to last element, and so on. Try accessing z[-4] and see what happens.

What if you wanted to get the first and second elements of z and not the last one. You can do this using index ranges:

```
>>> z[0:2]
(5, 0)
```

How about the second and the third:

```
>>> z[1:3]
(0, 34.34)
```

Let's make a bigger tuple so that we can do some more indexing:

```
>>> w = (1, 2, 3, 4, 5, 6, 7, 8, 9, 10)
```

This has 10 elements:

```
>>> len(w)
10
```

To get the first five, you can do:

```
>>> w[0:5]
(1, 2, 3, 4, 5)
```

Now, if the first index in your range happens to be zero, then you can skip it. So, instead of w[0:5], you can just write:

```
>>> w[:5]
(1, 2, 3, 4, 5)
```

Let's get the last 3:

```
>>> w[7:10]
(8, 9, 10)
```

Note that this gives you w[7], w[8], w[9]. But you can get the same thing by not specifying the last index:

```
>>> w[7:]
(8, 9, 10)
```

So, when you ask for $w[i{:}j]$ you get all elements between $w[i]$ and $w[j-1]$. If you don't specify the first index, it defaults to 0. If you don't specify the last index, it defaults to the length of the tuple.

Let's try one more thing:

```
>>> w[4:5]
(5,)
```

This gave you a tuple with just the fifth element. Look at how a tuple with a single element is defined. It is (5,) instead of (5). The comma is essential! This is because without the comma (5) is just the number five. With the comma, it is a tuple with one element — the number five. See the difference:

```
>>> type((5))
<class 'int'>
```

This is an integer.

```
>>> type((5,))
<class 'tuple'>
```

This is a tuple with one element.

How can I have a tuple with zero elements? Can you guess? Well, it is just ():

```
>>> ()
()
```

Here is another way to get a tuple with zero elements:

```
>>> w[5:5]
()
```

```
>>> len(w[5:5])
0
```

Alright, now some more advanced indexing. You do not have to remember everything. I am just putting them here to see what is possible. First, you can have ranges with negative numbers. For example, this is how to get everything from the fourth element counting from the end to the second element counting from the end:

```
>>> w[-4:-2]
(7, 8)
```

And if you want to go all the way to the last element you can write:

```
>>> w[-4:10]
(7, 8, 9, 10)
```

Or you can completely skip the last element:

```
>>> w[-4:]
(7, 8, 9, 10)
```

Meditate a little bit with this example with negative indices:

```
>>> w[-4:-1]
(7, 8, 9)
```

This means that `w[-i:-j]` gives everything between `w[len(w)-i]` and `w[len(w)-j]` for i and j positive numbers in this example.

What happens if you try to give a range that doesn't make sense. Like, take me all the way from the fifth element to the second element:

```
>>> w[5:2]
()
```

You get an empty tuple. Nice! What happens if you mix positive and negative indices:

```
>>> w[-6:8]
(5, 6, 7, 8)
```

It works just fine if the ranges make sense![2]

This indexing business is getting quite long, but let me give you one more example before we move on. Here is how you can skip elements in between:

```
>>> w[0:10:2]
(1, 3, 5, 7, 9)
>>> w[1:5:2]
(2, 4)
```

So, `w[i:j:k]` will give you every `k`th element from `i` (inclusive) to `j` (exclusive). For example, if you want everything, you can do this:

```
>>> w[1:5:1]
(2, 3, 4, 5)
```

Let's say now that we want to take every other element as we did above. We can do this:

```
>>> w[0:10:2]
(1, 3, 5, 7, 9)
```

But because we start at zero, we can skip the first index:

```
>>> w[:10:2]
(1, 3, 5, 7, 9)
```

And because we end at `len(w)` (which is 10), we can also skip the second index:

```
>>> w[::2]
(1, 3, 5, 7, 9)
```

This is neat! Similarly, if we want to take every third element we do:

```
>>> w[::3]
(1, 4, 7, 10)
```

---

[2]Of course, avoid these indexing ranges in practice. I can guarantee that you will not remember what on earth you are doing if you look at the code the next day after you wrote it!

You can also make `k` negative. Let's see what happens:

```
>>> w[::-1]
(10, 9, 8, 7, 6, 5, 4, 3, 2, 1)
```

Wow, the elements are reversed! How about this:

```
>>> w[::-2]
(10, 8, 6, 4, 2)
```

Reversed and we skip every other element! Note that `w[::-2]` is equivalent to:

```
>>> w[10:0:-2]
(10, 8, 6, 4, 2)
```

So, the range has to make sense. You say, I want to go from the tenth element to the first one backwards skipping every other one.

2.1.2 *Concatenating tuples*

You can concatenate, i.e., join, tuples. Here is how:

```
>>> x + y
(1, 2, 2.1232, 3.14)
>>> x + y + z
(1, 2, 2.1232, 3.14, 5, 0, 34.34)
```

Nice![3] Observe that `x + y` and `y + x` are different:

```
>>> x + y
(1, 2, 2.1232, 3.14)
>>> y + x
(2.1232, 3.14, 1, 2)
```

---

[3]Wait a second! Why do we use the addition operator + to concatenate tuples? Isn't addition for numbers? Well, the meaning of the + operator depends on the type of the objects on which it is applied. If the objects are tuples, then it concatenates them. If the objects are numbers, then it adds them. This is called *operator overloading*. It helps us write more general and elegant code.

What if you add the () (0-element tuple) to a tuple:

```
>>> x + ()
(1, 2)
>>> () + x
(1, 2)
```

You just get the tuple back.[4]

2.1.3   *You cannot change tuples*

You cannot change a tuple once you create it. We say that tuples are *immutable*. For example, consider the previously created tuple w. You cannot do this:

```
>>> w[4] = 5
...
TypeError: 'tuple' object does not support item assignment
```

You get a TypeError.

Read the error message! It is quite clear. So, you should not be using tuple's for things that may change. You should only use tuple's for things that do not change. As a matter of fact, you should *always* use tuple's for things that do not change. In this way, if you try to change that object, you will get this nice error above instead of a nasty bug. If you want something that can be changed, then you need to use a list.

2.2   Python Lists: Colab

Lists are like tuples, containers of many things, but they can be changed. Here is a list:

```
>>> a = [1, 2, 3, 4, 5, 6, 7]
```

---

[4]Okay, okay! There is something deeper here. We just saw that + is an operator that takes two tuples and gives us two tuples. We also saw that there is a special tuple () that does not change anything when we add it to another tuple. There is one more property that holds for tuple: (x + y) + z is the same as x + (y+z). We say that tuple concatenation is *associative*. This seems to be by design! Mathematical constructs satisfying all these properties are called *monoids*. Tuples are a monoid!

Note that you define it with [] instead of (). Just like tuples, you can get the length of the list with `len`:

```
>>> len(a)
7
```

And you can index lists in exactly the same way you index tuples:

```
>>> a[0]
1
>>> a[1]
2
>>> a[2:5]
[3, 4, 5]
>>> a[::2]
[1, 3, 5, 7]
>>> a[::-2]
[7, 5, 3, 1]
```

You can also combine two lists into a new list, like this:

```
>>> b = [3.1, 5.0, 2.]
>>> c = a + b
>>> c
[1, 2, 3, 4, 5, 6, 7, 3.1, 5.0, 2.0]
```

And here is the empty list:

```
>>> []
[]
>>> type([])
<class 'list'>
>>> len([])
0
>>> a + []
[1, 2, 3, 4, 5, 6, 7]
```

So, lists behave just like tuples, but with one main difference: *you can change them!* For example, you can change the second element of the list a to 3.4:

```
>>> a
[1, 2, 3, 4, 5, 6, 7]
>>> a[1] = 3.4
```

```
>>> a
[1, 3.4, 3, 4, 5, 6, 7]
```

You can also put a new item at the end of the list using `list.append()`. Here is how:

```
>>> a
[1, 3.4, 3, 4, 5, 6, 7]
>>> a.append(6.3)
>>> a
[1, 3.4, 3, 4, 5, 6, 7, 6.3]
```

You can also remove the last item of the list using `list.pop()`:

```
>>> a
[1, 3.4, 3, 4, 5, 6, 7, 6.3]
>>> a.pop()
6.3
```

Notice that the item was returned by `pop()`. Here is the list after the pop operation:

```
>>> a
[1, 3.4, 3, 4, 5, 6, 7]
```

The last item was removed. Now `list.append()` and `list.pop()` are methods of the *class* `list`. We have not talked yet about classes. We will do it in a later chapter. For now, think of a class as an object that may have some data inside it as well as some functions that you can apply to it. The data and the functions of the class are called *attributes* and *methods*, respectively. When you ask for `help()` on a class, you will get a list of attributes and methods of the class. Try it on `list`. If you are using a Jupyter notebook, you can type the name of the object, followed by a `.`. Depending on where you run the notebook, you may immediately see the available methods or you may have to press `tab` on your keyboard.

Now back to lists. We saw `list.append()` and `list.pop()`. There are more built in methods. `list.insert()` is a useful one:

```
>>> a
[1, 3.4, 3, 4, 5, 6, 7]
>>> a.insert(4, 23.5325)
```

```
>>> a
[1, 3.4, 3, 4, 23.5325, 5, 6, 7]
```

Here is what it does:

```
>>> help(a.insert)
Help on built-in function insert:

insert(self, index, object, /) unbound builtins.list method
    Insert object before index.
```

list.`remove()` is another useful method:

```
>>> a
[1, 3.4, 3, 4, 23.5325, 5, 6, 7]
>>> a.remove(3.4)
>>> a
[1, 3, 4, 23.5325, 5, 6, 7]
```

Here is what it does:

```
>>> help(a.remove)
Help on built-in function remove:

remove(value, /) method of builtins.list instance
    Remove first occurrence of value.

    Raises ValueError if the value is not present.
```

Let's try to remove something that is not there:

```
>>> a.remove(0)
Traceback (most recent call last):
  File "<stdin>", line 1, in <module>
ValueError: list.remove(x): x not in list
```

As I said before: Read the error messages!!!

The list.`extend()` method allows you to append an entire list to a given list:

```
>>> a
[1, 3, 4, 23.5325, 5, 6, 7]
>>> a.extend([0.3, 0.12, 123.])
```

```
>>> a
[1, 3, 4, 23.5325, 5, 6, 7, 0.3, 0.12, 123.0]
```

You can also sort a list using `list.sort()`:

```
>>> a.sort()
>>> a
[0.12, 0.3, 1, 3, 4, 5, 6, 7, 23.5325, 123.0]
```

Here is how you can count the occurrence of a specific element in a list:

```
>>> a.count(1)
1
>>> a.count(0.0)
0
```

And here is how you can find the index of a list element:

```
>>> idx = a.index(23.5325)
>>> idx
8
>>> a[idx]
23.5325
```

There are more things you can do with lists, but this should be enough to get you started.

2.3 Numerical Arrays: Colab

The ability to do basic things with numerical arrays, i.e., vectors and matrices of real numbers, is extremely important in data science applications because both data and model parameters are represented with numerical arrays. The field of mathematics dealing with such objects is called *linear algebra*. We do linear algebra in Python using the Numpy.[5] Numpy is a huge Python library and it has a lot of goodies. We are going to cover the very basics here. When there

---

[5]Pronounced as *Num-pie*. You can find the complete documentation at https:// numpy.org.

is something specific that you want to do, the best way to figure out how to do it is to Google it or ask a generative AI.

First, let's import numpy. We typically, do it by creating a short-cut called np:

```
>>> import numpy as np
```

In this way, you can access whatever is in numpy by going through np which is less characters. Let's create a 1D numpy array from a list:

```
>>> a = np.array([1.2, 2.0, 3.0, -1.0, 2.0])
>>> print(a)
[ 1.2  2.   3.  -1.   2. ]
```

This behaves just like a vector in Matlab. You can multiply with a number:

```
>>> 2.0 * a
array([ 2.4,  4. ,  6. , -2. ,  4. ])
>>> a * 3
array([ 3.6,  6. ,  9. , -3. ,  6. ])
```

You can add another vector (of the same size):

```
>>> b = np.array([1, 2, 3, 4, 5])
>>> c = a + b
>>> c
array([2.2, 4. , 6. , 3. , 7. ])
```

You can raise all elements to a power:

```
>>> a ** 3
array([ 1.728,  8.   , 27.   , -1.   ,  8.   ])
```

Or you can apply a standard function to all elements:

```
>>> np.exp(a)
array([ 3.32,  7.39, 20.09,  0.37,  7.39])
>>> np.cos(a)
array([ 0.36, -0.42, -0.99,  0.54, -0.42])
>>> np.sin(a)
array([ 0.93,  0.91,  0.14, -0.84,  0.91])
```

```
>>> np.tan(a)
array([ 2.57, -2.19, -0.14, -1.56, -2.19])
```

Note that the functions you can use are in `np`. Here is how you can get how many elements you have in the array:

```
>>> a.shape
(5,)
```

Also, to see if this is a 1D array (higher dimensional arrays, 2D, 3D, etc. are also possible), you can use this:

```
>>> a.ndim
1
```

You access the elements of the array just like the elements of a list:

```
>>> a[0]
1.2
>>> a[1]
2.0
>>> a[2:4]
array([ 3., -1.])
>>> a[::2]
array([1.2, 3. , 2. ])
>>> a[::-1]
array([ 2. , -1. ,  3. ,  2. ,  1.2])
```

You can find the minimum or the maximum of the array like this:

```
>>> a.min()
-1.0
>>> a.max()
3.0
```

Note that `a` is an object of a class. The class is called `np.ndarray`:

```
>>> type(a)
numpy.ndarray
```

and `min()` and `max()` are functions of the `np.ndarray` class. Another function of the class is the dot product:

```
>>> a.dot(b)
20.2
```

You can achieve the same thing with the matrix multiplication operator `@`:

```
>>> a @ b
20.2
```

There is also a submodule called `numpy.linalg` which has some linear algebra functions you can apply to arrays. For 1D arrays (aka vectors) the vector norm is a useful one:

```
>>> np.linalg.norm(a)
4.409081537009721
```

This calculates the Euclidean norm of the vector. If the vector is

$$\mathbf{a} = [a_1, a_2, \ldots, a_n],$$

then the Euclidean norm is given by

$$\|\mathbf{a}\| = \sqrt{a_1^2 + a_2^2 + \cdots + a_n^2}.$$

When $n = 2$ or $n = 3$, this quantity is what we know as the magnitude of the vector. For larger n, it is a generalization of the concept of the magnitude of a vector. Try typing `help(np.linalg.norm)` to see what more options you have.

Exercises

1. Write Python code to evaluate the angle between the two vectors:

$$\vec{r}_1 = 4\hat{\mathbf{i}} + 3.5\hat{\mathbf{j}} + 2.5\hat{\mathbf{k}},$$

and

$$\vec{r}_2 = 1.5\hat{\mathbf{i}} + 2.5\hat{\mathbf{j}}.$$

Hint: Use tuples to represent `r1` and `r2`.

2. Consider the tuple:

```
q = (3.0, 23.0, 1e-3, 12.242, 23, 41, 50, 30., 20.)
```

Write Python code to do the following:

(a) Get the length of q.
(b) Get all the elements of q from the second to the second to last.
(c) Get every other element of q from the second to the second to last.
(d) Reverse q.

3. Consider the list:

```
data = [23.0, 21.0, 23.0, 23.0, 23.84, 24.52]
```

Write Python code to do the following:

(a) Find how many occurrences of 23.0 there are in data.
(b) Find the mean of the data list, i.e., the sum of all elements divided by their number. *Hint*: Use the built-in python function `sum()`.
(c) Find the minimum and the maximum of data. *Hint*: Use the function `min()` for the minimum and `max()` for the maximum.

4. Consider the list:

```
data = [1, 4, 3, 10, 4, 3, 4, 4]
```

Write Python code to do the following:

(a) Add the element 6 at the end of `data`.
(b) Add the list `[0, -1, 3]` at the beginning of `data`.
(c) Find how many elements are in the `data`.
(d) Extract all `data` elements from the second to the second to last (inclusive).
(e) Extract all `data` elements from the second to the second to last (inclusive) skipping every other element.
(f) Sort the list.
(g) Find the minimum of the list.
(h) Find the average of the elements in the list.

5. Consider the two vectors:

$$\vec{r}_1 = 1.0\hat{i} - 2.0\hat{j} + 3.0\hat{k},$$

and

$$\vec{r}_2 = 2.0\hat{i} + 2.0\hat{j} - 3.0\hat{k}.$$

Write Python code using NumPy to do the following:

(a) Find a unit vector in the direction of $\vec{r}_1$.
(b) Find the angle between $\vec{r}_1$ and $\vec{r}_2$ in radians.
(c) Calculate the projection of $\vec{r}_2$ onto $\vec{r}_1$, i.e.,

$$\text{proj}_{\vec{r}_1}\vec{r}_2 = \frac{\vec{r}_1 \cdot \vec{r}_2}{\|\vec{r}_1\|^2}\vec{r}_1.$$

(d) Use `numpy.cross` to find the cross product between $\vec{r}_2$ and $\vec{r}_1$.
(e) Verify numerically that $\vec{r}_1 \cdot (\vec{r}_2 \times \vec{r}_1) = 0$.

6. Consider the following code that creates a random array[6]:

```
import numpy as np
np.random.seed(12345)
x = np.random.randn(100)
```

Write Python code to do the following:

(a) Find the minimum of x.
(b) Find the maximum of x.
(c) Use `x.shape` to get the number of elements of x as an integer.
(d) Calculate the mean of x by summing all elements and dividing by the number of elements.
(e) Compare your calculated mean with the result of `x.mean()`. Are they the same?
(f) Create a new array x2 containing the squared elements of x.

---

[6]We will talk about random numbers in a later chapter. For now, just run the code and see what array is created. Note that every time you run the code, you get the same array. This is because we fixed the random seed.

(g) Calculate the variance of x by:

    i. Finding the mean of x2.

    ii. Subtracting from the result of the previous step the square of the mean of x.[7]

(h) Compare your calculated variance with x.var(). Are they the same?

---

[7]We will discuss what the variance is and prove this formula in a later chapter. Now, I want you to develop some Python skills.

Chapter 3

Data Loading and Selection

In this chapter, I introduce matrices and comma-separated-values (csv) files, showing you how to load csv data. I introduce the Python data analysis library (pandas) and dataframes as tables, demonstrating how pandas loads various file formats like csv and excel. I show you how to select columns, summarize data, and clean missing values.

3.1   Matrices: Colab

Matrices are used to represent data tables. You can think of the columns as different data characteristics and the rows as different entries like this:

$$\text{data-matrix} = \begin{bmatrix} \text{stress}_1 & \text{strain}_1 & \text{temperature}_1 \\ \text{stress}_2 & \text{strain}_2 & \text{temperature}_2 \\ \text{stress}_3 & \text{strain}_3 & \text{temperature}_3 \\ \vdots & \vdots & \vdots \\ \text{stress}_n & \text{strain}_n & \text{temperature}_n \end{bmatrix}.$$

You can also use matrices together with vectors to represent linear systems of equations. For example, the following system of equations,

$$\begin{cases} 2x + 3y = 1, \\ 4x + 5y = 2, \end{cases}$$

can be written in matrix form as

$$\begin{bmatrix} 2 & 3 \\ 4 & 5 \end{bmatrix} \begin{bmatrix} x \\ y \end{bmatrix} = \begin{bmatrix} 1 \\ 2 \end{bmatrix}.$$

I will use a bold capital letter for a matrix, e.g., $\mathbf{A}$. I will write the matrix elements also with capital letters, but not bold and with a subscript that picks the row and column, e.g., A_{ij} picks the element of $\mathbf{A}$ in the ith row and jth column. For an $n \times m$ matrix, the number of rows is n and the number of columns is m and I may write

$$\mathbf{A} = \begin{bmatrix} A_{11} & A_{12} & \cdots & A_{1m} \\ A_{21} & A_{22} & \cdots & A_{2m} \\ \vdots & \vdots & \ddots & \vdots \\ A_{n1} & A_{n2} & \cdots & A_{nm} \end{bmatrix}$$

or just

$$\mathbf{A} = (A_{ij})$$

if there is no confusion about the size of the matrix.

Let's recall matrix multiplication. Take an $n \times m$ matrix $\mathbf{A}$ and multiply it by an $m \times p$ matrix $\mathbf{B}$ to get an $n \times p$ matrix $\mathbf{C}$:

$$\mathbf{C} = \mathbf{AB}.$$

The element C_{ij} of the matrix $\mathbf{C}$ is given by

$$C_{ij} = \sum_{k=1}^{m} A_{ik} B_{kj}.$$

I will use lowercase bold letters to denote vectors like $\mathbf{x}$. I will refer to the ith element of the vector $\mathbf{x}$ as x_i. Now, we can write a vector as

$$\mathbf{x} = (x_1, x_2, \ldots, x_m).$$

Alternatively, we can think of it as a column matrix:

$$\mathbf{x} = \begin{bmatrix} x_1 \\ x_2 \\ \vdots \\ x_n \end{bmatrix}.$$

Matrix vector multiplication is then straightforward. Take an $n \times m$ matrix $\mathbf{A}$ and multiply it by the m-dimensional vector $\mathbf{x}$ to get an

n-dimensional vector $\mathbf{y}$:

$$\mathbf{y} = \mathbf{A}\mathbf{x}.$$

The element y_i of the vector $\mathbf{y}$ is given by

$$y_i = \sum_{j=1}^{m} A_{ij} x_j.$$

Let's make a matrix to play with in Python[1]:

```
>>> import numpy as np
>>> np.random.seed(1234)
>>> A = np.random.rand(4, 6)
>>> A
array([[0.19, 0.62, 0.44, 0.79, 0.78, 0.27],
       [0.28, 0.80, 0.96, 0.88, 0.36, 0.50],
       [0.68, 0.71, 0.37, 0.56, 0.50, 0.01],
       [0.77, 0.88, 0.36, 0.62, 0.08, 0.37]])
```

Note that we used a list of rows and each row was a list of numbers. Here are the dimensions of A:

```
>>> A.ndim
2
>>> A.shape
(4, 6)
```

You see that A.shape has two entries. The first is the number of matrix rows and the second is the number of matrix columns. Let's access some elements of A:

```
>>> A[0, 0]
0.1915194503788923
>>> A[2, 2]
0.37025075479039493
```

---

[1]To save some space in the book, I am rounding the numbers to two decimal places. So, if you try the code yourself, the results may differ from what you see here.

You can access the 3rd row like this:

```
>>> A[2]
array([0.68346294, 0.71270203, 0.37025075, 0.56119619, 0.50308317,
       0.01376845])
```

Here is how you can get the third column:

```
>>> A[:, 2]
array([0.43772774, 0.95813935, 0.37025075, 0.36488598])
```

You can get a submatrix, say, the first 3×3 submatrix like this:

```
>>> A[:3, :3]
array([[0.19, 0.62, 0.44],
       [0.28, 0.80, 0.96],
       [0.68, 0.71, 0.37]])
```

and so on.

You can multiply matrices with numbers:

```
>>> 2.0 * A
array([[0.38, 1.24, 0.88, 1.57, 1.56, 0.54],
       ...
       [1.55, 1.77, 0.73, 1.23, 0.15, 0.74]])
```

```
>>> A * 3.0
array([[0.57, 1.87, 1.31, 2.36, 2.34, 0.82],
       ...
       [2.32, 2.65, 1.09, 1.85, 0.23, 1.11]])
```

You can raise them to a power:

```
>>> A ** 3
array([[0.01, 0.24, 0.08, 0.48, 0.47, 0.02],
       ...
       [0.46, 0.69, 0.05, 0.23, 0.00, 0.05]])
```

You can pass them through functions:

```
>>> np.exp(A)
array([[1.21, 1.86, 1.55, 2.19, 2.18, 1.31],
       ...
       [2.17, 2.42, 1.44, 1.85, 1.08, 1.45]])
```

You can add them together:

```
>>> A + A
array([[0.38, 1.24, 0.88, 1.57, 1.56, 0.55],
       ...
       [1.55, 1.77, 0.73, 1.23, 0.15, 0.74]])
```

You can multiply them together (assuming that the dimensions match):

```
>>> B = np.random.randn(A.shape[1], 6)
>>> B
array([[ 0.19,  0.55,  1.32, -0.47,  0.68, -1.82],
       ...
       [-1.40, -0.10, -0.55, -0.14,  0.35, -0.04]])
```

```
>>> C = A.dot(B)
>>> C
array([[-0.13,  0.25,  0.53, -1.52,  2.10, -0.49],
       ...
       [-0.18,  0.93,  0.59, -1.50,  2.10, -0.28]])
```

```
>>> C.shape
(4, 6)
```

You can multiply a matrix with a vector:

```
>>> a = np.random.randn(A.shape[1])
>>> a
array([ 0.57,  1.55, -0.97, -0.07,  0.31, -0.21])
>>> A.dot(a)
array([0.77, 0.41, 1.24, 1.35])
```

You can also interchange the rows of a matrix with each column. Let's make a new matrix for this:

```
>>> A.T
array([[0.19, 0.28, 0.68, 0.77],
       ...
       [0.27, 0.50, 0.01, 0.37]])
```

```
>>> A.T.shape
(6, 4)
```

There are some special numpy functions for creating arrays. Here is how to make an array of zeros:

```
>>> np.zeros((4, 5))
array([[0., 0., 0., 0., 0.],
       [0., 0., 0., 0., 0.],
       [0., 0., 0., 0., 0.],
       [0., 0., 0., 0., 0.]])
```

An array of ones:

```
>>> np.ones((5, 2))
array([[1., 1.],
       [1., 1.],
       [1., 1.],
       [1., 1.],
       [1., 1.]])
```

A unit matrix:

```
>>> np.eye(5)
array([[1., 0., 0., 0., 0.],
       [0., 1., 0., 0., 0.],
       [0., 0., 1., 0., 0.],
       [0., 0., 0., 1., 0.],
       [0., 0., 0., 0., 1.]])
```

There are some useful functions that can be applied to matrices. For example, here is how you can sum all the elements of a matrix:

```
>>> A.sum()
12.78
```

But what if you wanted to sum all the rows for each column? You can do it like this:

```
>>> A.sum(axis=0)
array([1.92, 3.02, 2.13, 2.84, 1.72, 1.16])
```

Let's convince ourselves that this works as expected by working, say, on the third column of A. Here is the code to get the third row:

```
>>> A[:, 2]
array([0.44, 0.96, 0.37, 0.36])
Let's sum it up:
>>> A[:, 2].sum()
2.13
```

Is this the same as the third element of A.sum(axis=0)?

Okay, so when we do A.sum(axis=0), we are saying sum over axis 0 of the matrix, which corresponds to the rows. What if we wanted to sum over the columns: We could do this:

```
>>> A.sum(axis=1)
array([3.09, 3.77, 2.84, 3.08])
```

Verify this by hand as well by extracting the second row of the matrix and summing it up.

3.2 Comma-Separated-Values (csv) Files: Colab

How do we store data on a computer? There is a plethora of data storing formats. For example, Microsoft Excel has a format for storing data. There are also several high-performance binary data formats which are common, e.g., HDF5. However, by far, the most ubiquitous data format is the "comma-separated values" file format (csv).

The csv file format can be used to store a matrix. Here is how. First of all, the file is a text file. Each row of the file contains numbers separated by commas. There must be as many entries in a row as matrix columns. If you start a row with #, it will be ignored when reading the file.

Let's look at an example of such a file from a catalysis experiment (Katsounaros *et al.*, 2012). The file is called catalysis.csv.[2]

---

[2]You can download the data repository of this book.

The contents of the file are as follows:

```
Time,NO3,NO2,N2,NH3,N2O
0,500.00,0.00,0.00,0.00,0.00
30,250.95,107.32,18.51,3.33,4.98
60,123.66,132.33,74.85,7.34,20.14
90,84.47,98.81,166.19,13.14,42.10
120,30.24,38.74,249.78,19.54,55.98
150,27.94,10.42,292.32,24.07,60.65
180,13.54,6.11,309.50,27.26,62.54
```

Each row corresponds to a different time instant. Columns 1–5 give the mass in grams of a chemical species at that instant of the chemical reaction.

Let's use numpy to load the file. Here is how:

```
>>> import numpy as np
>>> data = np.loadtxt('catalysis.csv', delimiter=',')
>>> data
array([[   0.00, 500.00,   0.00,   0.00,   0.00,   0.00],
       [  30.00, 250.95, 107.32,  18.51,   3.33,   4.98],
       [  60.00, 123.66, 132.33,  74.85,   7.34,  20.14],
       [  90.00,  84.47,  98.81, 166.19,  13.14,  42.10],
       [ 120.00,  30.24,  38.74, 249.78,  19.54,  55.98],
       [ 150.00,  27.94,  10.42, 292.32,  24.07,  60.65],
       [ 180.00,  13.54,   6.11, 309.50,  27.26,  62.54]])
```

That's it. You can now do whatever you want with the data. For example, let's see if mass is conserved in this reaction. First, take everything except the first column (the first column is time, all the rest are masses):

```
>>> data[:, 1:]
array([[500.00,   0.00,   0.00,   0.00,   0.00],
       [250.95, 107.32,  18.51,   3.33,   4.98],
       [123.66, 132.33,  74.85,   7.34,  20.14],
       [ 84.47,  98.81, 166.19,  13.14,  42.10],
       [ 30.24,  38.74, 249.78,  19.54,  55.98],
       [ 27.94,  10.42, 292.32,  24.07,  60.65],
       [ 13.54,   6.11, 309.50,  27.26,  62.54]])
```

Now, let's sum all columns for each row:

```
>>> data[:, 1:].sum(axis=1)
array([500.00, 385.09, 358.32, 404.71, 394.28, 415.40, 418.95])
```

We see that the mass is 500 grams initially, and then it decreases. This means that there is one (or more) intermediate chemical species that have not been measured! You cannot violate the conservation of mass (at these energies).

Now, let me show you how simple it is to save data in csv files. First, let us create some synthetic data at random:

```
>>> np.random.seed(12345)
>>> X = np.random.randn(10, 3)
>>> X
array([[-0.20,  0.48, -0.52],
       [-0.56,  1.97,  1.39],
       [ 0.09,  0.28,  0.77],
       [ 1.25,  1.01, -1.30],
       [ 0.27,  0.23,  1.35],
       [ 0.89, -2.00, -0.37],
       [ 1.67, -0.44, -0.54],
       [ 0.48,  3.25, -1.02],
       [-0.58,  0.12,  0.30],
       [ 0.52,  0.00,  1.34]])
```

We have 10 rows and three columns (whatever they mean). To save them in a csv file, we do

```
>>> np.savetxt('random_data.csv', X, delimiter=',')
```

If you now look in your working directory, you will see the file `random_data.csv`.

3.3 Python Data Analysis Library (Pandas): Colab

Pandas[3] is a Python library that provides fast, flexible, and data structures that let you work real-life data quickly. Typically, we

---

[3]https://pandas.pydata.org/.

import the library like this:

```
>>> import pandas as pd
```

Let's load the `catalysis.csv` file we created before. We can now load the file into a pandas DataFrame:

```
>>> data = pd.read_csv('catalysis.csv')
>>> data
# Time      NO3      NO2       N2     NH3     N20
0       0  500.00     0.00     0.00    0.00    0.00
1      30  250.95   107.32    18.51    3.33    4.98
2      60  123.66   132.33    74.85    7.34   20.14
3      90   84.47    98.81   166.19   13.14   42.10
4     120   30.24    38.74   249.78   19.54   55.98
5     150   27.94    10.42   292.32   24.07   60.65
6     180   13.54     6.11   309.50   27.26   62.54
```

DataFrame is a two-dimensional data structure that can store data of different types (e.g., numbers, strings, booleans) in a table format. It is similar to a spreadsheet. The columns can have names. Here are the column names:

```
>>> data.columns
Index(['Time', 'NO3', 'NO2', 'N2', 'NH3', 'N20'], ...)
```

The function `DataFrame.describe()` provides a summary of the data. It shows the mean, standard deviation, minimum, maximum, and quartiles[4]:

```
>>> data.describe().round(2)
         Time      NO3      NO2       N2     NH3     N20
count    7.00     7.00     7.00     7.00    7.00    7.00
mean    90.00   147.26    56.25   158.74   13.53   35.20
std     64.81   175.83    55.22   129.63   10.50   26.63
min      0.00    13.54     0.00     0.00    0.00    0.00
25%     45.00    29.09     8.26    46.68    5.34   12.56
50%     90.00    84.47    38.74   166.19   13.14   42.10
```

---

[4]We will explain what quartiles are in a later chapter.

| 75% | 135.00 | 187.30 | 103.06 | 271.05 | 21.80 | 58.32 |
| max | 180.00 | 500.00 | 132.33 | 309.50 | 27.26 | 62.54 |

Dataframes allow you to access data by column names. For example, to get the NO3 column, you can do

```
>>> data['NO3']
0     500.00
1     250.95
2     123.66
3      84.47
4      30.24
5      27.94
6      13.54
Name: NO3, dtype: float64
```

What type of object is this? Let's check:

```
>>> type(data['NO3'])
```

Aha! It's a pandas series! You can think of these as a column of a table.

What if you want to extract the data pertaining to NO3 and NO2? You can do this by passing a list of column names to the dataframe:

```
>>> data[['NO3', 'NO2']]
        NO3       NO2
0    500.00      0.00
1    250.95    107.32
2    123.66    132.33
3     84.47     98.81
4     30.24     38.74
5     27.94     10.42
6     13.54      6.11
```

The type of the object is a dataframe. What if I ask for a column that does not exist? Try it out to see what kind of error you get.

Another useful operation is to extract a subset of the dataframe based on some condition. For example, let's extract the data for the first 90 minutes:

```
>>> data[data['Time'] <= 90]
     Time      NO3      NO2       N2     NH3     N2O
0       0   500.00     0.00     0.00    0.00    0.00
1      30   250.95   107.32    18.51    3.33    4.98
2      60   123.66   132.33    74.85    7.34   20.14
3      90    84.47    98.81   166.19   13.14   42.10
```

What is going on here? What on earth is `data['Time'] <= 90`? Let's print it to see:

```
>>> data['Time'] <= 90
0      True
1      True
2      True
3      True
4      False
5      False
6      False
Name:  Time, dtype: bool
```

It looks like a Boolean `Series` object. It contains `True` and `False` values. When you pass it to the dataframe, it only keeps the rows where the condition is `True`. This is neat!

Let's do an OR operation. For example, let's get the rows where the time is either less than 60 minutes or greater than 150 minutes:

```
>>> data[(data['Time'] < 60) | (data['Time'] > 150)]
     Time      NO3      NO2       N2     NH3     N2O
0       0   500.00     0.00     0.00    0.00    0.00
1      30   250.95   107.32    18.51    3.33    4.98
6     180    13.54     6.11   309.50   27.26   62.54
```

The | symbol stands for the logical OR operator. It is applied elementwise to the `Series` objects. If it sees a `True` or a `False`, it returns `True`. Only if it sees two `False`, it returns `False`.

How do we do a NOT operation? The operator we need is \~. For example, let's get the rows where the time is not between 60 and 150 minutes:

```
>>> data[~(data['Time'] >= 60) & (data['Time'] <= 150)]
    Time      NO3      NO2      N2    NH3    N2O
0      0   500.00     0.00    0.00   0.00   0.00
1     30   250.95   107.32   18.51   3.33   4.98
```

Let's repeat the total mass calculation that we did before:

```
>>> total_mass = data.iloc[:, 1:].sum(axis=1)
>>> total_mass
0     500.00         '
1     385.09
2     358.32
3     404.71
4     394.28
5     415.40
6     418.95
dtype: float64
```

As we discussed before, there may be a species that we are not taking into account. Let's call that species x. The mass of that species is the total mass minus the mass of the other species:

```
>>> mass_X = 500 - data.iloc[:, 1:].sum(axis=1)
>>> mass_X
0       0.00
1     114.91
2     141.68
3      95.29
4     105.72
5      84.60
6      81.05
dtype: float64
```

Let's add a column to the dataframe:

```
>>> data['X'] = mass_X
>>> data
```

	Time	NO3	NO2	N2	NH3	N2O	X
0	0	500.00	0.00	0.00	0.00	0.00	0.00
1	30	250.95	107.32	18.51	3.33	4.98	114.91
2	60	123.66	132.33	74.85	7.34	20.14	141.68
3	90	84.47	98.81	166.19	13.14	42.10	95.29
4	120	30.24	38.74	249.78	19.54	55.98	105.72
5	150	27.94	10.42	292.32	24.07	60.65	84.60
6	180	13.54	6.11	309.50	27.26	62.54	81.05

Let's save our work to a new file:

```
>>> data.to_csv('catalysis-with-X.csv')
```

If you are working on Google Colab, download the file to your local machine. If you are working on your own computer, locate the file, and open it with Excel.

Exercises

1. Write Python code that constructs the 90-degree rotation matrix in 2D:

$$A = \begin{bmatrix} \cos(\pi/2) & \sin(\pi/2) \\ -\sin(\pi/2) & \cos(\pi/2) \end{bmatrix}.$$

Then:

(a) Verify by direct computation that the matrix A rotates the vector $x = (1,0)$ by 90 degrees counterclockwise. That is, $Ax = (0,1)$.

(b) Create the matrix:

$$B = A \cdot A.$$

(c) Is B a rotation matrix? What is the effect of B on the vector x? *Hint*: To verify that B is a rotation matrix, you need to check that $B^T \cdot B = I$ (i.e., that B is orthogonal) and that the determinant of B is 1. You can use the "np.linalg.det()" function to compute the determinant of a matrix.

(d) Apply the matrix B to the vector x and verify that it rotates the vector by 180 degrees counterclockwise.

(e) If you perform 90 degree rotation four times, what is the effect of the resulting matrix on the vector x? Verify your answer by showing that multiplying A by itself four times results in the identity matrix.

2. In this problem, you are going to use the `pandas` library to analyze a real dataset collected as part of the NSF-funded project. You can find the raw dataset in the data repository of this book. The dataset contains temperature data from a residential community in Indiana. Download it and open it up in Excel to explore it. Then go over the following steps:

(a) Use the `pd.read_excel` function to read the `temperature_raw.xlsx` file. Name the dataframe you read `df`.

(b) Print the first 10 rows of the dataframe `df` using the `df.head()` method.

(c) Print the last 10 rows of the dataframe `df` using the `df.tail()` method.

(d) Print summary statistics of the dataframe `df` using the `df.describe()` method.

(e) Note that there is a column called `date`, but `pandas` is not aware that this is actually a date. Let's make it aware of that using the `pd.to_datetime()` function. Run the following code:

```
df.date = pd.to_datetime(df['date'], format='%Y-%m-%d')
```

(f) This is real data. There are some missing values. Let's fill them with the mean value of the column. Missing values are represented as `NaN` (Not a Number). Count how many `NaN` values you have in each column. *Hint*: Use the `df.isna()` method to create a Boolean array that is True for `NaN` values and False otherwise. Then, use the `np.sum()` function to count the number of True values. True is equivalent to 1 and False is equivalent to 0.

(g) Clean the dataset by dropping all `NaN` values. Call the cleaned data `df\_clean`. Use the `df.dropna()` method.

(h) Verify that there are no `NaN` values in `df_clean`.

(i) How many unique households do we have and what are their unique names? *Hint*: Use the `df.unique()` method.

(j) Print summary statistics for your cleaned dataset.

(k) Save your cleaned dataset in a csv file using `df_clean.to_csv()`. Use the name `temperature_clean.csv`.

(l) Save the cleaned data set in a Microsoft Excel format. Look at `pandas.DataFrame.to_excel` for the right function. Call the file `temperature_clean.xlsx`.

(m) Download the newly created `temperature_clean.xlsx` file to your computer and open it up. There is no need to show something here. Just do it for your own education. If you are running the Jupyter notebook on your own computer, then there is nothing to do apart from finding the file. If you are working on Google Colab, see this stackoverflow post.

Chapter 4

Data Visualization

In this chapter, I introduce you to the matplotlib Python library, which is essential for data visualization. I demonstrate how to create line plots and scatter plots, which are fundamental visualization techniques. Then, I explain what a histogram is and show you how to plot one effectively. Finally, I discuss methods for comparing two histograms, allowing you to better understand the relationships between different data distributions.

4.1 Plotting Simple Functions: Colab

We plot using the matplotlib Python library. Plotting is a huge topic and we cannot cover everything here. In general, if you know the type of plot that you want to do, a simple Google search of the type "how to X using matplotlib" will probably send you to an example that you can adjust to your needs.

Here is how to import `matplotlib`:

```
import matplotlib.pyplot as plt
```

Let's start by plotting simple 1D functions. We are going to plot this function:

$$f(t) = e^{-0.1t}[\cos(\pi t) + 0.5\sin(\pi t)],$$

for t between 0 and 4π. You may have guessed it, but you just plotted the position of a harmonic oscillator with damping! Think of the response of a spring-mass system with some dashpot taking out energy (e.g., the suspension system of car).

First, you generate the data that you want to plot. Here is a dense set of t's and the $f(t)$'s[1]:

```
import numpy as np
ts = np.linspace(0, 4 * np.pi, 100)
ys = np.exp(-0.1 * ts) * (np.cos(np.pi * ts) \
    + 0.5 * np.cos(np.pi * ts))
```

Now, you can plot the function like this:

```
plt.plot(ts, ys);
plt.show()
```

The `plt.show()` function is not needed in Jupyter notebooks. In a Jupyter notebook, the plot will be displayed inline. But you are running the code within a Python script or a Python shell; you need to call `plt.show()` and it will open a new window with the plot. This is going to be the simplest possible plot you can make. There is no title, no labels, no legend, no nothing. I'll show you how to add all these things gradually.

First, instead of the simple `plt.plot(ts, ys)`, we get an explicit axis handle and use it to add the title, labels, etc. It works like this:

```
fig, ax = plt.subplots()
ax.plot(ts, ys)
ax.set_xlabel('$t$')
ax.set_ylabel('$y$')
ax.set_title('Some title')
plt.show()
```

We replaced `plt.plot(ts, ys)` with `ax.plot(ts, ys)`. And we set the axis labels and title using the `set_xlabel`, `set_ylabel`, and `set_title` methods of the axis handle.

Let's now add one more function to the plot. I am going to add the derivative of the position as a function of time, i.e., the velocity

---

[1]The backslash (\) at the end of the first line in the following code block is used to break a long line of code into multiple lines for better readability.

of the harmonic oscillator. It is

$$v(t) = f'(t)$$
$$= -0.1e^{-0.1t}[\cos(\pi t) + 0.5\sin(\pi t)]$$
$$+ e^{-0.1t}[-\pi\sin(\pi t) + 0.5\pi\cos(\pi t)]$$
$$= -0.1f(t) + e^{-0.1t}[-\pi\sin(\pi t) + 0.5\pi\cos(\pi t)].$$

Like this:

```
vs = -0.1 * ys \
    + np.exp(-0.1 * ts) * (-np.pi * np.sin(np.pi * ts) \
    + 0.5 * np.pi * np.cos(np.pi * ts))
fig, ax = plt.subplots()
ax.plot(ts, ys, label='$f(t)$')
ax.plot(ts, vs, '--', label='$v(t)$')
ax.set_xlabel('$t$')
ax.set_ylabel('$y$')
ax.set_title('Some title')
ax.legend(loc='best')
plt.show()
```

You can see the figure we just made in Figure 4.1. There is a lot to unpack in the code above. Note that I have used a dashed line for the velocity plot. This is done by passing the string '--' to the plot function. There are many different styles of lines you can use and they are named similarly to the ones in MATLAB, e.g., ':' is a dotted line and '-.' is a dash-dot line. We could also specify the color of the line as in MATLAB, e.g., 'r--' is a red dashed line. You can experiment with the styles in the Jupyter notebook of this section.

I have also added a legend. This is done by calling the legend method of the axis handle. The loc='best' argument tells the legend to be placed in the best location. This is usually a place in the plot that doesn't overlap a lot with the data. It's a good default. The label arguments of the plot function are used to label the curves in the legend. The text of the labels can be whatever you want. In engineering applications, you often want to use mathematical notation

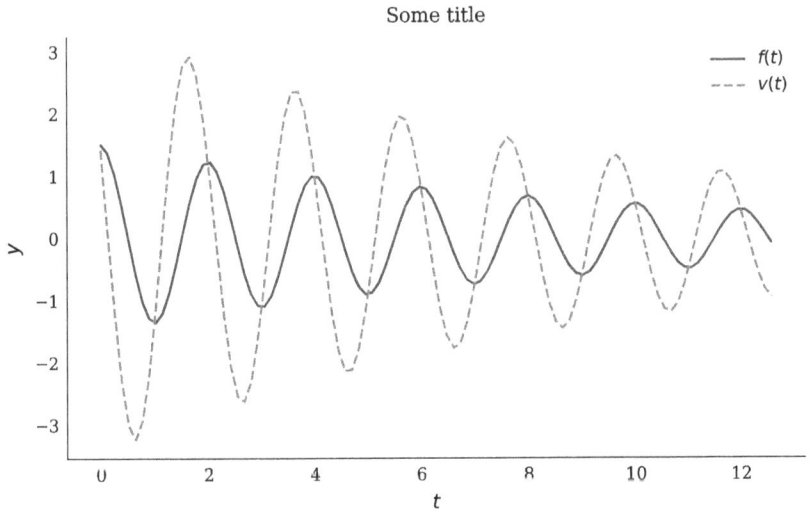

Figure 4.1. A simple plot with two curves made with `matplotlib`. The data mirror the ideal behavior of a harmonic oscillator with damping. The solid line is the position, and the dashed line is the velocity.

in the labels. You do this by putting the formula in dollar signs. The language we use to do this is called LaTeX.[2]

Let's plot the same data on the so-called *phase space* (position vs velocity):

```
fig, ax = plt.subplots()
ax.plot(ys, vs)
ax.set_xlabel('$f(t)$ (position)')
ax.set_ylabel('$v(t)$ (velocity)')
plt.show()
```

You can see the figure we just made in Figure 4.2. We get a nice spiral centered around $(0,0)$. For the damped harmonic oscillator, $(0,0)$ is the steady state solution. This is where the physical system goes after a lot of time.

---

[2] As a matter of fact, this whole book is written in LaTeX. You will see a lot of LaTeX in the Jupyter notebooks that accompany this book. Click on any of the equations to see the code that generates them. Eventually, you can start picking up the syntax by yourself.

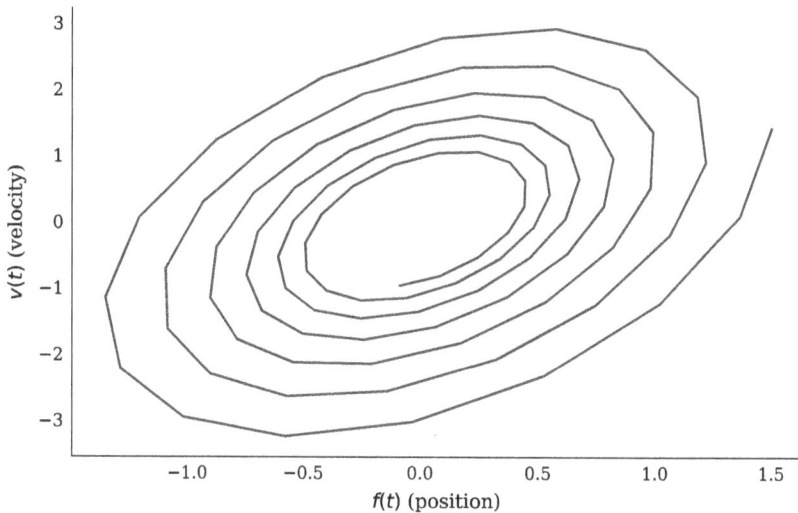

Figure 4.2. The trajectory in the phase space of the harmonic oscillator with damping.

4.1.1 Styling your plots

If you simply run the codes above, the style of your plots will be different from mine. The colors will be slightly different, the fonts will be different, and there will be a bounding box around the plot. In the figures that I create professionally, I am trying to adhere to the data–ink ratio maximization principle of the work of Tufte (2001). The data–ink ratio is the ratio of the area of the pixels used to display the data to the area of the pixels used to display the axes, labels, and titles. Tufte argues that the data-pixel ratio should be as high as possible. In other words, we should use as much of the space as possible to display the data. You will have to look at the Jupyter notebook details to see how I am achieving this. It is not, however, essential if all you want is to learn how to plot.

4.1.2 Saving figures to files

You now know how to make figures. You can see them in the Jupyter notebooks you are running or you can display them in a separate window. But what if you want to extract them so that you can put them in a paper or your presentation? You can take a screenshot of them, but that's kind of ridiculous, isn't it? Here is how you can save a figure to pretty much any format you like:

```
fig.savefig('pendulum.png')
```

This is saved in your current working directory. And you can save your figure in multiple formats:

```
fig.savefig('pendulum.jpg')
fig.savefig('pendulum.pdf')
fig.savefig('pendulum.svg')
```

4.2   Plotting Noisy Measurements: Colab

Let's continue on the previous example with the damped harmonic oscillator by making things a bit more realistic. Now, assume that we have a small error in the measurement of the position. So, we assume that our measurement y_i at timestep t_i is not exactly given by the function $f(t_i) = e^{-0.1t_i} \left[\cos(\pi t_i) + 0.5 \sin(\pi t_i) \right]$ but by

$$y_i = f(t_i) + \text{noise}.$$

This is a model of the *measurement process*. The measurement process is one of the fundamental blocks in data science problems. We do not know enough yet to understand where the noise comes from. Instead, I am going to just make up some noise for you and tell you how you can make it bigger or smaller. Then, we are going to simulate the measurement process and generate some data to plot. To proceed on your own, you need

```
import numpy as np
import matplotlib.pyplot as plt
```

These are the times at which I will have measurements:

```
ts = np.linspace(0, 4 * np.pi, 100)
```

This is the true (but unobserved) oscillator position:

```
fs = np.exp(-0.1 * ts) * (np.cos(np.pi * ts) \
    + 0.5 * np.cos(np.pi * ts))
```

And let's make the some noise!

This is the variable that controls the noise:

```
sigma = 0.1
```

It is called *standard deviation*. We explain what it is in a few chapters. For now, this is what you need to know:

- Smaller standard deviation means less noise.
- Bigger standard deviation means more noise.
- The smallest standard deviation you can have is zero (no noise).

And now we are going to make some random numbers centered at zero, and we are going to multiply them with `sigma` and add them to `fs`:

```
ys = fs + sigma * np.random.randn(fs.shape[0])
```

The noisy position measurements are in `ys`. Let's plot them against the true (but, I repeat, *unobserved*) oscillator positions:

```
fig, ax = plt.subplots()
ax.plot(ts, fs, label='Unobserved true position')
ax.plot(ts, ys, '.', label='Noisy measurement of position')
ax.set_xlabel('$t$ (Time)')
ax.set_ylabel('$y$ (Position)')
plt.legend(loc='best')
```

You can see the figure we just made in Figure 4.3.

Typically, we do not observe the velocity of the oscillator. We can, however, reconstruct the velocity from the position measurements. This is done by estimating the derivative via *finite differences*, i.e., by

$$v(t) \approx \frac{y(t + \delta t) - y(t)}{\delta t},$$

for δt being the time that passes between two consecutive timesteps. The Numpy library has a function for reconstructing the velocity in this way. It is called `np.gradient`. Let's use it to reconstruct the velocity. We have to use the noisy position data:

```
dt = ts[1] - ts[0]
vs = np.gradient(ys, dt)
```

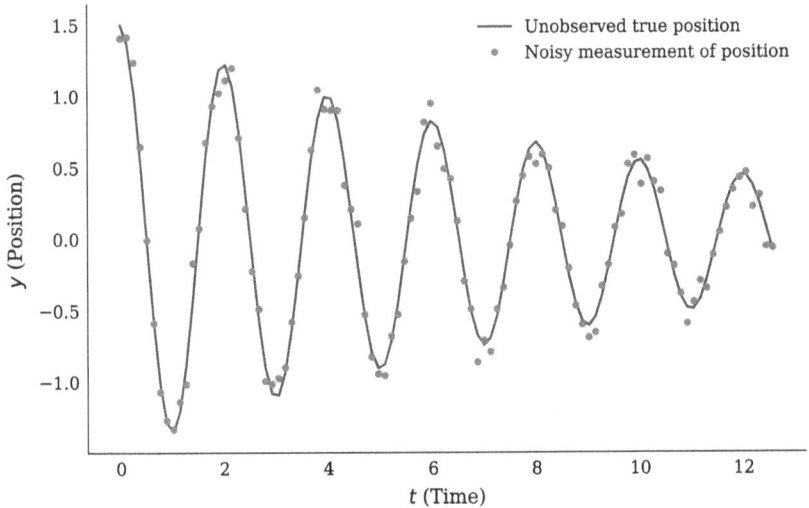

Figure 4.3. The oscillator position with measurement noise.

Now, we can plot the reconstructed velocity against the true velocity (see Figure 4.4):

```
vs_true = -0.1 * fs + np.exp(-0.1 * ts) \
    * (-np.pi * np.sin(np.pi * ts) \
    + 0.5 * np.pi * np.cos(np.pi * ts))
fig, ax = plt.subplots()
ax.plot(ts, vs_true,
    label='Unobserved true velocity')
ax.plot(ts, vs, '.',
    label='Reconstruction of velocity from noisy measurements')
ax.set_xlabel('$t$ (Time)')
ax.set_ylabel('$v$ (Velocity)')
ax.legend(loc='best')
```

Observe that the noise of the velocity is bigger than that of the position. With the things that we learn in this class, it is actually possible to predict how much uncertainty there is going to be in the estimate of the velocity. Using state-of-the-art data science, it is also possible to construct a much more accurate estimate the velocity than

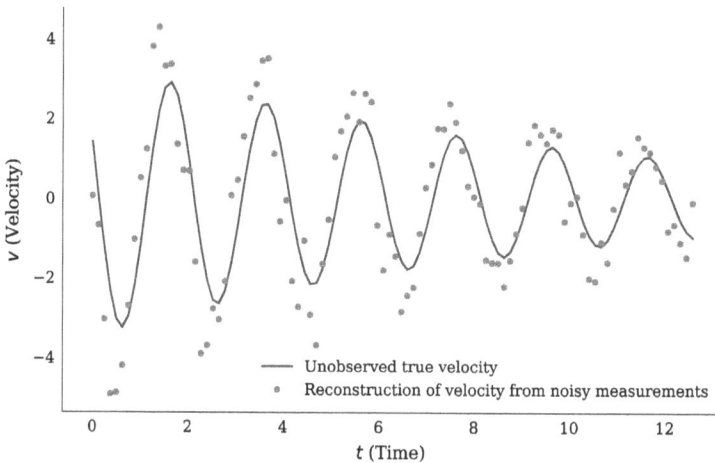

Figure 4.4. The oscillator velocity estimated from noisy positions.

the naive method we used above. However, these so-called *filtering problems* are outside of the scope of this book.[3]

4.3   Scatter Plots: Colab

Scatter plots are a nice way to investigate if there are correlations between measured scalar variables. I am going to use the `temp_price.csv` dataset, which you can download from the data repository of this book. The dataset is a single day version of the `temperature_raw.xlsx` dataset from Exercise 2 of Chapter 3. First, make sure that the data are in your working directory. Then, let's read the data, drop all the rows that have missing values, and rename the columns:

```
>>> import pandas as pd
>>> data = pd.read_csv('temp_price.csv')
>>> clean_data = data.dropna(axis=0)
>>> clean_data = clean_data.rename(
    columns={'Price per week': 'week_price',
             'Price per day': 'daily_price'})
```

---

[3]If you are interested in this topic, you need to learn about *Kalman filters*.

```
>>> clean_data.head().round(2)
    household    date         score    t_out    ...
0   a1           2019-01-06   85       38.6     ...
1   a10          2019-01-06   70       38.6     ...
2   a11          2019-01-06   61       38.6     ...
3   a12          2019-01-06   65       38.6     ...
4   a13          2019-01-06   66       38.6     ...
```

If you look closely at the data entries, you will note that the dataset
includes only one day, 01/06/2019, and that the average temperature
is 38.6°F. We have 50 different records, each one corresponding to a
different apartment in the same residential building.

Let us see first how energy consumption hvac correlates with the
weekly bill week_price:

```
fig, ax = plt.subplots()
ax.scatter(clean_data['week_price'], clean_data['hvac'])
ax.set_xlabel('Weekly price')
ax.set_ylabel('Energy consumption')
```

You can see the figure we just made in Figure 4.5. This kind of plot
is called a *scatter plot*. Does Figure 4.5 make sense? The more energy

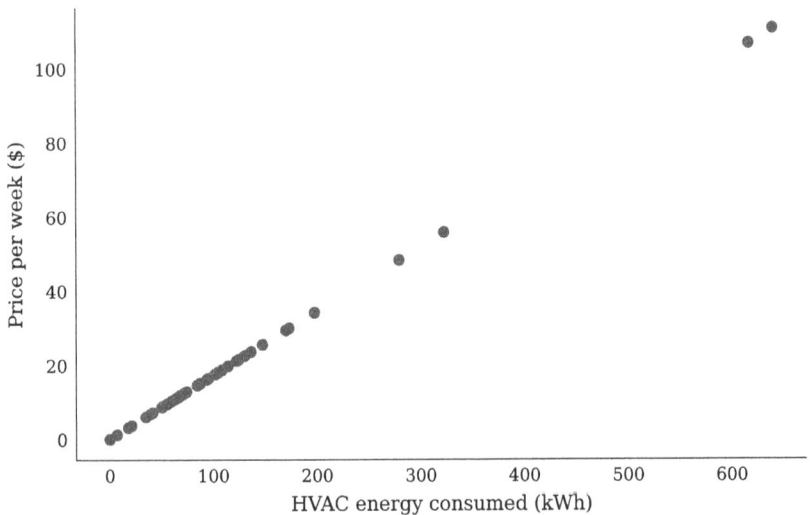

Figure 4.5.   Scatter plot of energy consumption vs weekly price.

energy a unit consumes the higher the energy bill. The relationship between the two is linear reflecting the fact that each household is paying for the same price per kWh.

The relationship depicted here is also *causal*. The energy consumption is causing the bill to be higher. However, this causality direction does not come from the data alone. It comes from us using our knowledge about how energy bills are calculated.

What other causal relationships could we have? Let's look at the temperature of each household `t_unit` vs the energy consumed `hvac`. Here is the scatter plot:

```
fig, ax = plt.subplots()
ax.scatter(clean_data['t_unit'], clean_data['hvac'])
ax.set_xlabel('Temperature')
ax.set_ylabel('Energy consumption')
```

You can see the figure we just made in Figure 4.6. We observe that higher unit temperature, in general, leads to higher HVAC energy consumption. However, the relation is not one-to-one. This is because the apartments in this building have different physical characteristics. For example, an apartment that is at the corner of the building has more of its external surfaces exposed to the environment and thus it needs more energy to maintain a given temperature than an

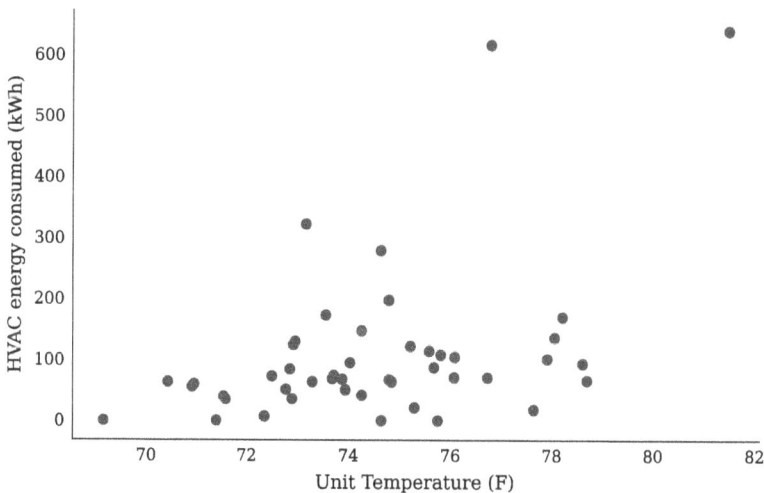

Figure 4.6. Scatter plot of energy consumption vs temperature.

apartment that is, say, in the middle of the building. As a matter of fact, note that there are some apartments that consume zero energy even though the external temperature is 38°F. These people are getting their heating for free from their neighbors!

So, the relationship between unit temperature and HVAC energy consumption is causal to some degree. But there are other variables that affect it as well. Here, these are the physical characteristics of the apartment and the temperature of the neighboring units. More often than not, this is the situation we find ourselves when dealing with real datasets.

4.4   Histograms: Colab

Histograms offer a nice way to summarize the uncertainty (or variability) of scalar variables. I am assuming that you have seen histograms in the past. They work by splitting the interval in which your variable takes values into bins and counting how many times the variable falls inside each bin.

Let's look at some examples. We start with the histogram of the apartment temperature `t_unit` in the `clean_data` dataset. Here is the code that produces the histogram in Figure 4.7:

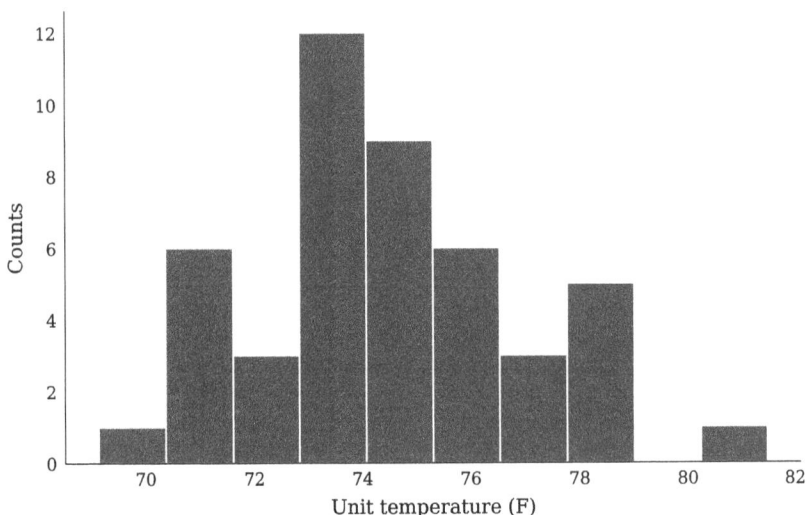

Figure 4.7.   Histogram of the apartment temperature.

```
fig, ax = plt.subplots()
ax.hist(clean_data['t_unit'])
ax.set_xlabel('Unit temperature (F)')
ax.set_ylabel('Counts')
```

It is straightforward to read this. Each bar gives you the number of households with internal temperature that fall in the bin.

Sometimes, we want to normalize the height of the bars so that the total area covered by the histogram is one. To do this, you need to divide by the total number of observations and by the width of each bin. What we get is a density. We say that the histogram is *normalized*. We see in later lectures that this is an approximation of a probability density of a random variable. To get the density, you need to use the keyword `density=True` in `hist`. Here is how:

```
ax.hist(clean_data['t_unit'], density=True)
```

You can see the figure we just made in Figure 4.8. You can also change the bin number. The default is 10. Let's make it 5:

```
ax.hist(clean_data['t_unit'], bins=5)
```

You can see the figure we just made in Figure 4.9.

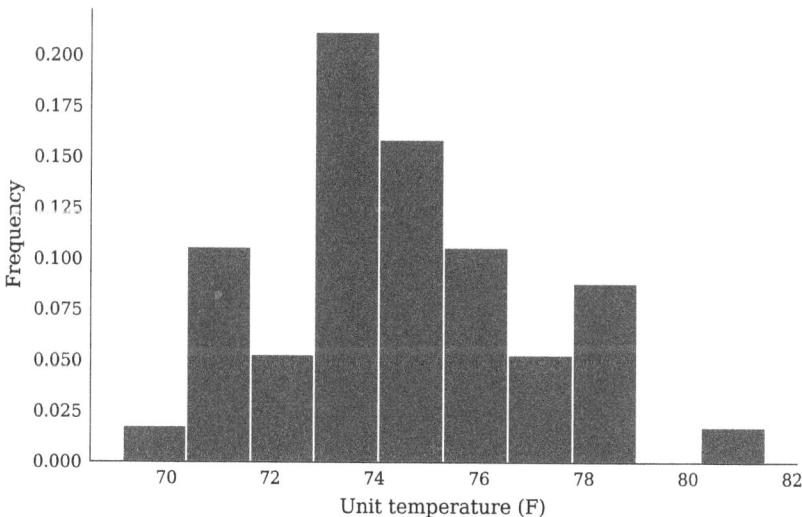

Figure 4.8. Normalized histogram of the apartment temperature.

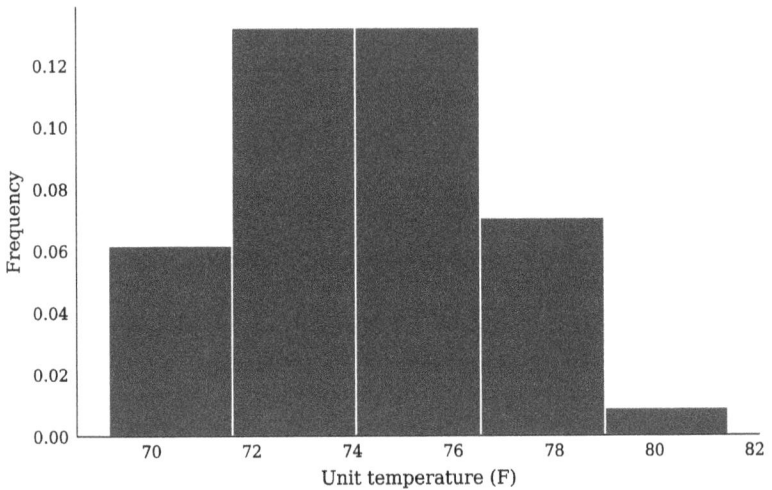

Figure 4.9. Histogram of the apartment temperature with five bins.

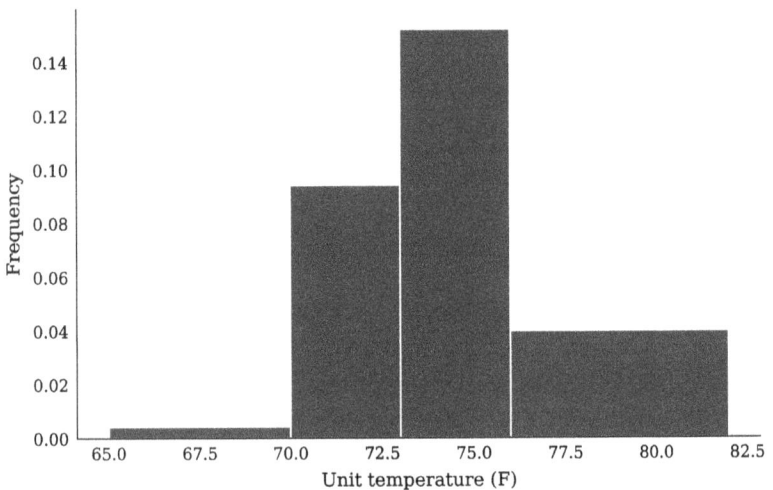

Figure 4.10. Histogram of the apartment temperature with specified bin edges.

Alternatively, you can also specify the bins on your own. You just have to provide the bin edges. Let's pick: $(65, 72, 76, 82)$. Here, we go:

```
ax.hist(clean_data['t_unit'], density=True, bins=(65, 70, 73, 76,
↪  82))
```

You can see the figure we just made in Figure 4.10.

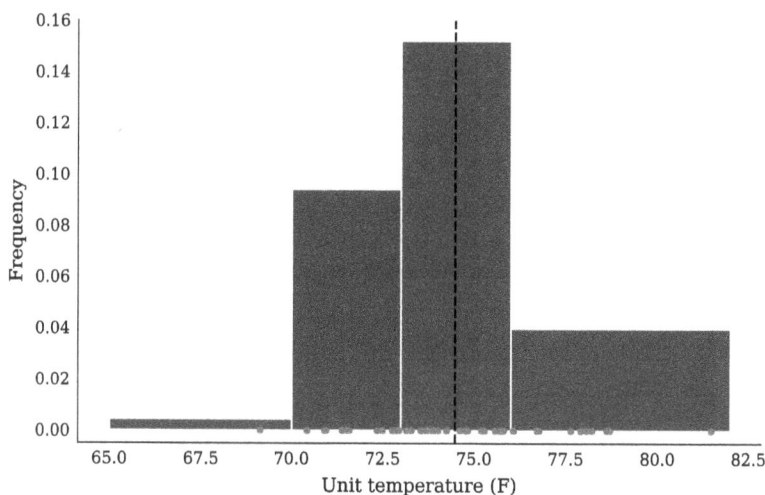

Figure 4.11. Histogram of the apartment temperature with raw data points and mean temperature.

Let's plot a few more things on our histogram. For example, let's plot the raw data as points on the x-axis:

```
ax.plot(clean_data['t_unit'], np.zeros(clean_data.shape[0]), '.')
```

Next, let's add also the mean temperature as a vertical line:

```
ax.axvline(clean_data['t_unit'].mean(), color='k', linestyle='--')
```

You can see the figure we just made in Figure 4.11.

Exercises

1. Remake Figures 4.1 and 4.2 using a higher data collection frequency. We used 100 timesteps between times 0 and 4π. Double and quadruple that number to make the plotted curves look smoother.
2. Remake Figure 4.2 for a much longer time so that your spiral really starts touching $(0, 0)$.
3. Remake Figure 4.4 with smaller and smaller noise standard deviation σ until you get an accurate estimate of the velocity.

4. Using the `clean_data` dataset of Section 4.3, do the scatter plot between unit temperature `t_unit` and the variable `score`. The variable `score` is a measure we developed in our NSF project to characterize how well occupants behave related to energy consumption. It is a number between 0 and 100. The bigger the score, the better the household is at conserving energy. Note that we calculate this score using the detailed behavior of the household during an entire week. We look at things like the temperature setpoint the occupants pick when they are home during the day, when they sleep, and when they are away. We do not just look at the unit temperature. Answer the following questions:

   (a) Is there a correlation between `t_unit` and `score`?
   (b) Based on what I told you about the calculation of the `score`, is there a causal relationship between the variables?

5. Remake a plot like Figure 4.11 for the `score` variable.
6. **Visual analysis of a variable-speed compressor experiment:** In this problem, you are going to need this dataset. The dataset was kindly provided by Prof. Ziviani. You can either download manually and put it in your working directory or just download it with `curl` (if you are on a Mac, Linux, or Google Colab). The data are part of an experimental study of a variable speed reciprocating compressor. The experimentalists varied two temperatures T_e and T_c (both in degrees C) and they measured various other quantities. Your goal is to understand the experimental design and develop some intuition of the map between T_e and T_c and measured capacity and power (both in W). Answer the following questions:

   (a) Do the scatter plot of T_e and T_c. This will reveal the experimental design picked by the experimentalists. Make sure you label the axes correctly. Is there a gap in the experimental design? If yes, why do you think they have a gap? *Hint*: These are columns `T_e` and `T_c` of the dataframe `data`.
   (b) Create scatter plots to explore the relationships between variables: (1) `T_e` vs `Capacity`, (2) `T_c` vs `Capacity`, (3) `T_e` vs `Power`, (4) `T_c` vs `Power`, (5) `T_e` vs `T_c`, and (6) `T_c` vs `T_e`.
   (c) We are lucky that we only have two experimental control variables because can do a bit more thing with scatter.

You can color each point in the scatter plot according to a scale that follows an output variable. Let me show you what I mean by doing the plot for the `Capacity`:

```
from matplotlib import cm
fig, ax = plt.subplots()
cs = ax.scatter(data['T_e'], data['T_c'],
                c=data['Capacity'],
                cmap=cm.jet)
```

Now, repeat the same thing for the `Power` variable.

7. **Visual analysis of an airfoil experiment:** In this problem, you are going to repeat what you did in the previous problem but without my guidance! The dataset you are going to use is the Airfoil Self-Noise Data Set. This dataset from NASA contains measurements of NACA 0012 airfoils of varying sizes tested at different wind tunnel speeds and angles of attack. Throughout all experiments, both the airfoil span and observer position remained constant. The dataset includes the following input variables: "frequency, in Hertz," "angle of attack, in degrees," "chord length, in meters," "free-stream velocity, in meters per second," "suction side displacement thickness, in meters," and a single output variable "scaled sound pressure level, in decibels." You can load the data like this:

```
raw_data = np.loadtxt('airfoil_self_noise.dat')
df = pd.DataFrame(raw_data,
    columns=['Frequency', 'Angle_of_attack', 'Chord_length',
             'Velocity', 'Suction_thickness', 'Sound_
             ↪ pressure'])
```

(a) Do the histograms of all variables. Plot the histogram of each variable in a different plot. Make sure you label the axes correctly.

(b) Do the scatter plot between all input variables. This will give you an idea of the range of experimental conditions. Are there any holes in the experimental dataset, i.e., places where you have no data?

(c) Do the scatter plot between each input variable and the output. This will give you an idea of the relationship between each input and the output. Do you observe any obvious patterns?

(d) Now, pick the two input variables you think are the most important and do the scatter plot between them using the output to color the points; see the last question of the previous problem. Feel free to repeat it with more than two pairs of inputs if you want. Briefly discuss your findings.

Chapter 5

Printing, Functions, Data Manipulation, and Models

In this chapter, you learn the basics of string manipulation and formatting in Python. You discover how to implement and document Python functions, enabling you to write reusable and maintainable code. The chapter also covers techniques for applying functions to dataframe columns, for transforming and analyzing your data efficiently. Finally, you explore the fundamental concept of a model as a function.

5.1   Basics of Strings: Colab

You will often have to write some text to the screen. Python `str`'s ("str" stands for "string") are used to represent text. Here is a string:

```
>>> x = "This is a string"
>>> x
'This is a string'
```

We could have also used single quotes:

```
>>> x = 'This is the same string'
>>> x
'This is the same string'
```

Here is the string type:

```
>>> type(x)
<class 'str'>
```

You can do many things with strings. For a complete list, see `help(str)`. Here, we will survey the most frequently encountered string operations.

First, you can join two strings together:

```
>>> y = ', and another string.'
>>> y
', and another string.'
>>> z = x + y
>>> z
'This is the same string, and another string.'
```

Here is how you can center a string:

```
>>> z.center(60)
'          This is the same string, and another string.          '
```

You can also multiply a string with a number to repeat it many times:

```
>>> '=' * 60
'============================================================'
```

A string is essentially a list of characters. You can ask how many characters are in a string:

```
>>> len(z)
44
```

You can also make substrings using indices. These work in exactly the same way as the do for tuples and lists:

```
>>> z[::-1]
'.gnirts rehtona dna ,gnirts emas eht si sihT'
```

Here is how you can get a substring:

```
>>> z[2:10]
'is is th'
```

Here is how you can get a character:

```
>>> z[0]
'T'
```

Another useful thing is padding a string representing a number with zeros. This is particularly useful when reading many data files numbered as "001, 002, 003," etc. Here it is:

```
>>> '23'.zfill(5)
'00023'
>>> '1'.zfill(4)
'0001'
```

There is also an empty string:

```
>>> ''
''
>>> len('')
0
>>> ''.zfill(4)
'0000'
```

Now I am going to show you the most important thing you need to remember: How to turn integers and floats into strings so that you present the result of your analysis. Let's get some numbers first:

```
>>> a = 123 # an integer
>>> b = 12.908450 # a floating point number
```

Let's say that you want to put these numbers into a string so that you print them on the screen or maybe write something in a text file. Let's keep it simple. Say that we want to write: "I found out that a=[replace with value for a] and that b=[replace with value for b]." Here is one way to do this:

```
>>> x = 'I found out that a=' + str(a) + ' and that b=' + str(b)
>>> x
'I found out that a=123 and that b=12.90845'
```

So, all we did is turn the numbers into strings (using str(number)) and then added the strings together. This is not the best way to do it because it doesn't allow us to change the number of significant digits we present. The best way to do it is to use string formatting:

```
>>> x = f'I found out that a={a:d} and that b={b:1.2f}'
>>> x
'I found out that a=123 and that b=12.91'
```

Let me explain what this does. First, notice the "f" at the beginning of the string, which indicates this is an f-string (formatted string literal). Inside the f-string, we have expressions in curly brackets: `a:d` and `b:1.2f`. These expressions are evaluated at runtime. The variable name before the colon (e.g., `a` in `a:d`) refers directly to the variable we want to include in the string. The characters after the colon in the brackets tell Python how you would like to turn that variable into a string. The `d` means that you are expecting a decimal integer. The `1.2f` means that you are expecting a floating point number and that you want to keep two significant digits.

Here are some other examples of formatting:

```
>>> f'This is a={12:5d} with five characters in total padding with
↪    empty space'
'This is a=   12 with five characters in total padding with empty
↪    space'
>>> import math
>>> f'This is pi={math.pi:1.30f} (thirty digits of pi)'
'This is pi=3.141592653589793115997963468544 (thirty digits of pi)'
>>> f'This is b={b:1.3e} in scientific notation with three
↪    significant digits'
'This is b=1.291e+01 in scientific notation with three significant
↪    digits'
>>> f'This is a={a:o} in octal format.'
'This is a=173 in octal format.'
>>> f'And this is a={a:x} in hex format'
'And this is a=7b in hex format'
```

You can read more about formatted string literals.

5.2 The Print Function: Colab

So far, we were only able to show things in the screen if we did something like this:

```
>>> x = 'This is a string'
>>> x
'This is a string'
```

This doesn't work in an actual Python program, however. It only works in the Python shell or a Jupyter notebook. But even in Jupyter notebooks, if you wanted to show more than one variable, this would not work. Try this on a Jupyter notebook:

```
y = 'This is another string'
x
y
```

The output is:

```
'This is another string'
```

Only y was displayed. To display x and then y, you need to use the print() function. Here is how you do it:

```
print(x)
print(y)
```

The output is:

```
This is a string
This is another string
```

Just like formatting, there are a lot of details in print(). Here, we will be only covering the very basics. Let's start by printing some numerical variables:

```
a = 123
b = 98.2082
print(a)
print(b)
```

The output is:

```
123
98.2082
```

You can also print the variables one after the other:

```
print(a, b)
```

The output is:

```
123 98.2082
```

Notice that `print()` adds an empty space between the arguments that you want to print. There is no limit to how many arguments `print()` can have. The following also works:

```
print(a, b, x, y, 'and some more')
```

The output is:

```
123 98.2082 This is a string This is another string
  and some more
```

Of course, when you print numerics, you probably want to do it like this:

```
print(f'a={a:d}, b={b:1.2f}')
```

The output is:

```
a=123, b=98.21
```

That is, instead of just printing `a` and then `b`, you make a string with just the desired format and you print that string.

Notice that `print()` puts a newline each time it is called. See this:

```
print('This is the first line.')
print('This will appear in the second line.')
```

The output is:

```
This is the first line.
This will appear in the second line.
```

There is a way, a special character called `\n` which can be used to mark a newline in a string. When `print()` sees this character, then it adds a newline. See how it works:

```
print('This is the first line.\nThis will appear in the second
 ↪  line.')
```

The output is:

```
This is the first line.
This will appear in the second line.
```

Characters like \n are called special characters. In this class, we will use the \t character very often to align what we print. This is called the tab character. Here is what it does:

```
print('A\tB\tC')
```

The output is:

```
A    B    C
```

Get it? Here is why it is useful. Let's print a table of some fake data:

```
print('ID\tTime (s)\tMass (kg)\t')
print('-' * 40) # Prints 40 '-'
print(f'{0:d}\t{0.01:1.2f}\t\t{12.2:1.2f}')
print(f'{1:d}\t{0.02:1.2f}\t\t{12.5:1.2f}')
print(f'{2:d}\t{0.05:1.2f}\t\t{13.1:1.2f}')
print(f'{3:d}\t{0.07:1.2f}\t\t{13.6:1.2f}')
print(f'{4:d}\t{0.08:1.2f}\t\t{14.2:1.2f}')
```

The output is:

```
ID  Time (s)    Mass (kg)
----------------------------------------
0   0.01        12.20
1   0.02        12.50
2   0.05        13.10
3   0.07        13.60
4   0.08        14.20
```

Notice that I had to use two \t's to align the mass. It's probably too much to try to remember all these details. We will see many more examples as we go along and you will pick up certain tricks.

5.3   Python Functions: Colab

The function objects in Python allow you to give a name to a series of expressions that may or may not return a result. We have already seen a few functions. `print()` is a function, `type()` is a function, **help()** is a function. We have also used some standard mathematical functions like **math.exp()**. Functions are extremely important in data science. The models we build are essentially functions with parameters that are tuned to the data. So, we have to understand how functions work.

The syntax for defining a function in Python is as follows:

```
def function_name(some_inputs):
    some expressions
    return something # optional
```

Keep in mind that function inputs are also called *arguments*. Let's start with very simple functions.

5.3.1   *Functions with no inputs that return nothing*

First, let's look at functions that have no inputs and return nothing. Here is an example:

```
def print_hello():
    """Prints hello.
    """
    print('Hello there!')
```

Let's run this:

```
>>> print_hello()
Hello there!
```

See for yourself that the type of `print_hello` is a function:

```
>>> type(print_hello)
<class 'function'>
```

The other thing that I want you to notice is the text in the triple quotes below the function definition. This is called the *docstring* of the function. You use it to document what the function does so that other people know how to use it. It is not necessary to have a doc-string. However, it is a very good practice to do so if you want to remember what your code does. The docstring is what the `help()` function sees. Check this out:

```
>>> help(print_hello)
Help on function print_hello in module __main__:

print_hello()
    Prints hello.
```

Finally, let's see if `print_hello()` returns anything. Let's try to grab whatever it returns and print it.

```
>>> res = print_hello()
>>> print('Res is: ', res)
Hello there!
Res is:  None
```

Alright! Now you see why None is useful. A function that returns nothing, actually returns None.

Let me end this section by saying a few things about how I chose the function name. I could have called the function `f()` or `ph()` or `banana()`. Why did I choose the name `print_hello()`? Well, because this is what the function does. It prints "hello". In general, it is a good idea to pick nice descriptive names for your functions. Since the functions typically do something, you should use a verb in the function's name. Also, use complete words. Do not abbreviate the words. Write `print_hello()` instead of `prnt_hello()`. Notice also that I am using underscore to separate the words. Do that as well. It's a style preference, but it makes your code looks better. Also, do not use any capital letters unless it is absolutely justified. For example, do not use `Print_hello()` or `print_Hello()` or `Print_Hello()`. You will not remember which one you picked and you would have to go back and check wasting your time. Using only lower case letters and underscores makes it easier to remember the function names.

5.3.2 *Functions with inputs that return nothing*

Let's make a function that takes the name of someone as input and prints a hello statement. Here you go:

```
def print_hello_to(name):
    """Prints hello using the name provided.

    Arguments:

    name     -    The name of a person.

    Returns: Nothing
    """
    print('Hello there ' + name + '!')
```

Let's use it:

```
>>> print_hello_to('Ilias')
Hello there Ilias!
>>> print_hello_to('Philippos')
Hello there Philippos!
```

Now, see the `help()` function applied to `print_hello_to`:

```
>>> help(print_hello_to)
Help on function print_hello_to in module __main__:

print_hello_to(name)
    Prints hello using the name provided.

    Arguments:

    name     -    The name of a person.

    Returns: Nothing
```

You can have multiple inputs to a function. Let's write a function with two inputs:

```
def print_hello_to_two(name1, name2):
    """Prints hello using the names provided.

    Arguments:

    name1    -    The name of the first person.
```

```
name2     -     The name of the second person.

Returns: Nothing
"""
print('Hello there ' + name1 + ' and ' + name2 + '!')
```

Let's use it:

```
>>> print_hello_to_two('Ilias', 'Philippos')
Hello there Ilias and Philippos!
```

5.3.3   *Numerical functions*

When I am talking about numerical functions, I mean things like `math.exp()` or `math.log()` or `math.sqrt()` or `math.sin()` and so on. These functions typically take a single input that is a real number and they return also a single input which is also a real number. Here are some examples:

```
def square(x):
    """Calculates the square of ``x``.

    Arguments:
    x      -     The real number you wish to square

    Returns: The square of ``x``.
    """
    return x ** 2
```

Let's use it:

```
>>> help(square)
Help on function square in module __main__:

square(x)
    Calculates the square of ``x``.

    Arguments:
    x      -     The real number you wish to square

    Returns: The square of ``x``.
```

```
>>> square(2)
4
>>> square(23.0)
529.0
```

Because real functions are used so often, there is actually a shortcut. It is called *lambda functions*. To define a lambda function, the syntax is:

```
func_name = lambda inputs: single_expression_you_want_to_return
```

Here is the square function defined as a lambda function:

```
alt_square = lambda x: x ** 2
```

Let's use it:

```
>>> alt_square(2)
4
>>> alt_square(23.0)
529.0
```

You will see me using regular function definitions and lambda functions.

5.3.4   *What happens if you pass an input of the wrong type?*

Python does not check what types of inputs you give to a function. For example, you can pass to `square()` whatever type you want. Python will try to evaluate it and it will throw an error if it cannot do. Let's give `square()` a string to see what happens:

```
>>> square('one')
...
TypeError: unsupported operand type(s) for ** or pow(): 'str' and
↪  'int'
```

That's a very nice error message. It is a `TypeError`. You are going to feel the urge to skip over these messages. Don't do it! Read the message. It tells you where the problem is. The problem here is that the exponentiation operator `**` is not defined for strings. Let's replicate the error directly:

```
>>> 'one' ** 2
...
TypeError: unsupported operand type(s) for ** or pow(): 'str' and
↪    'int'
```

So, this kind of expression is meaningless in Python. Be careful with the inputs you pass to your functions.

5.3.5   *Default parameters*

Some times you may have a function with a lot of inputs. It is very annoying to have to provide a value for each input. If some of the inputs are used very often, you can provide a default value. The syntax for achieving this is as follows:

```
def function_name(input1, input2=default_value,
↪    input3=default_value, ...):
        ...
```

Let's demonstrate this by making a function that makes a plot of a numerical function on a given range. To try this out, you need to import `matplotlib.pyplot` and `numpy`. Here is the function definition:

```
def plot_func(f, left=0, right=1, style='-'):
    """Plot function f over the range of values [left, right] using
↪    color.

    Arguments:
    f      -   The function you want to plot. Yes, you can have a
↪    function
               as in input to another function!
    left   -   The left side of the interval over which you plot.
    right  -   The right side of the interval over which you plot.
    style  -   The style you want to use.
    """
    fig, ax = plt.subplots()
    xs = np.linspace(left, right, 100)
    ax.plot(xs, f(xs), style)
    ax.set_xlabel('$x$')
```

```
ax.set_ylabel('$f(x)$')
return fig, ax
```

There are two weird things here. First, notice that the input `f` is a function! Yes, you can have a function as an input to another function. No problem! Second, notice that the inputs are `left` and `right`, and `style` have *default values*. What does this mean? It means that if you do not provide a value for these inputs, then Python will use the default value. Such inputs are called *keyword arguments*.

Let's see how it works. The command:

```
plot_func(square)
```

will plot the square function over the range of the default values `[0, 1]` using the default style `'-'` (which is a solid line). See Figure 5.1. The following code will plot a cube in the default range and style:

```
cube = lambda x: x ** 3
fig, ax = plot_func(cube)
```

See Figure 5.2. But this code will plot a cube in the range `[-1, 1]` with a dashed line:

```
fig, ax = plot_func(cube, left=-1, style='--')
```

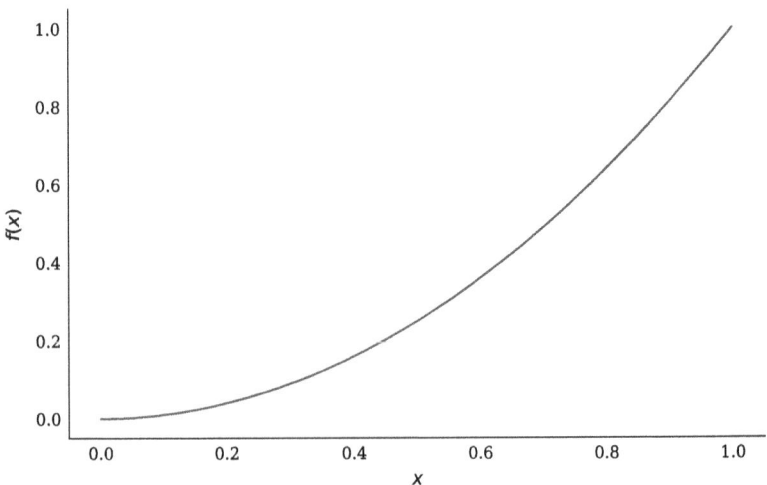

Figure 5.1. The square function using the default values.

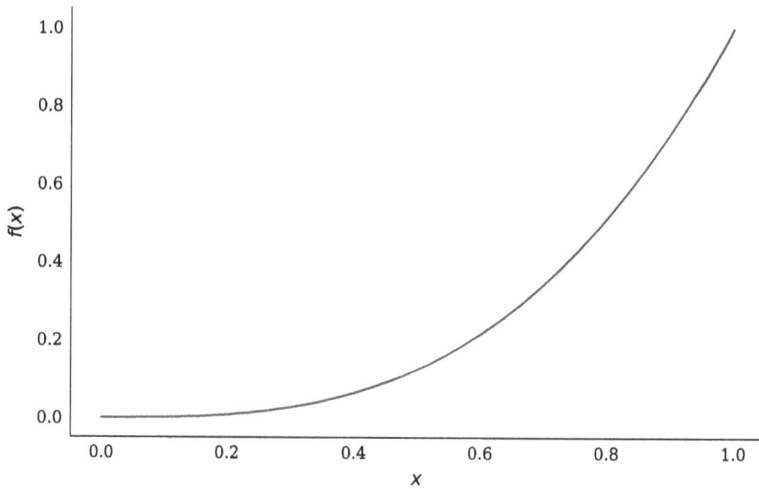

Figure 5.2.   The cube function.

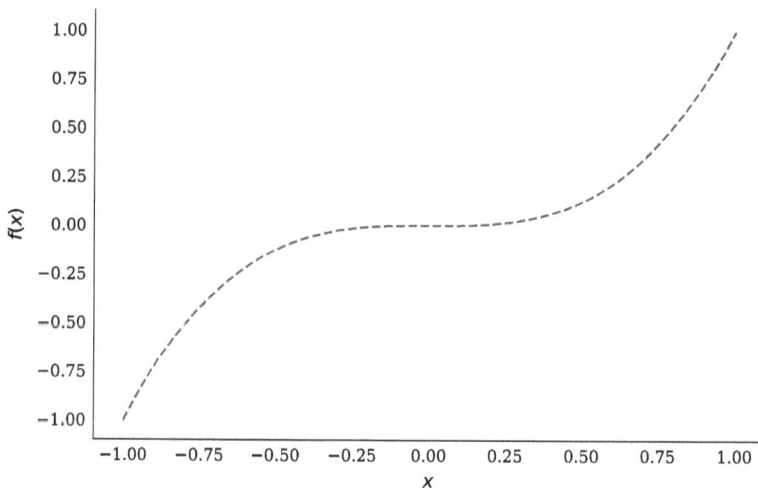

Figure 5.3.   The cube function in the range [-1, 1].

See Figure 5.3. Notice that the order of the keyword arguments does not matter. The following produces the same result:

```
fig, ax = plot_func(cube, left=-1, style='--')
```

But you cannot provide a keyword argument before a regular one. The following results in a SyntaxError:

```
fig, ax = plot_func(left=-1, style='--', cube)
```

Again, do not panic when you see an error like this! Read it. Doesn't it make sense? So, remember that regular arguments (otherwise known as positional arguments) must be before the keyword arguments.

5.3.6 *Applying functions to dataframes: Colab*

Let's reuse the `temp_price.csv` dataset we introduced in Section 3.3. Load it and clean it up as we did before, and name the resulting dataframe `clean_data`.

We are going to change units of temperature from degrees F to degrees C. We will use the function `pandas.DataFrame.apply()` to achieve this. First, we need to define the function that changes from degrees F to degrees C:

```
def change_degrees_F_to_C(deg_F):
    """Changes the temperature from degrees F to degrees C.

    Arguments:
    deg_F - The temperature in degrees F.

    Returns: The temperature in degrees C.
    """
    return (deg_F - 32) * 5 / 9
```

Now, we can update the two columns `t_out` and `t_unit`. This is a very elegant way to do it:

```
clean_data[['t_out', 't_unit']].apply(change_degrees_F_to_C)
```

Here is the result:

```
t_out    t_unit
0   3.66624 21.989280
1   3.66624 22.936811
...
```

In this way, we applied the function to each entry of the `clean_data[['t_out', 't_unit']]`. However, this change was not made to the `clean_data` dataframe. To make the change stick to the old data frame you need to do it like this:

```
clean_data_in_deg_C = clean_data.copy()
clean_data_in_deg_C[['t_out', 't_unit']] = \
    clean_data[['t_out', 't_unit']].apply(change_degrees_F_to_C)
clean_data_in_deg_C.head()
```

5.4   Models Are Functions: Colab

Let's now make our first model. A model is always some kind of function. It takes some input variables an it maps them to an output quantity of interest. A good model is one in which the input variables are the causes of the output quantity of interest. Let's make our first model for the `temperature_raw.xlsx` dataset of Exercise 2 of Chapter 2. Load it and prepare it like this:

```
df = pd.read_excel('temperature_raw.xlsx')
df = df.dropna(axis=0)
```

The column names are: `household`, `date`, `score`, `t_out`, `t_unit`, and `hvac`.

The output we want to look at is `hvac`, the weekly energy consumption of a unit in kWh. What affects the weekly energy consumption? Well, obviously the average external temperature `t_out`, which is given in degrees F. Let's do the scatter plot of the two variables to see what kind of relationship we have:

```
fig, ax = plt.subplots()
ax.scatter(df['t_out'], df['hvac'])
ax.set_xlabel('t_out (F)')
ax.set_ylabel('hvac (kWh)')
```

See Figure 5.4. Okay, it's kind of a mess, but we see four important trends:

- The energy consumption goes up when the average external temperature gets very cold. This makes sense because for very cold external temperatures the occupants are running their heating.

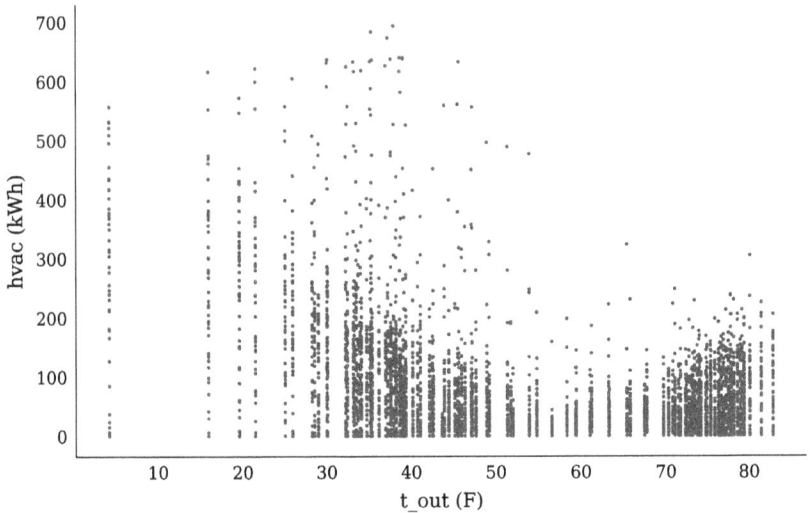

Figure 5.4.   The scatter plot of the weekly energy consumption `hvac` vs the external temperature `t_out`.

- The energy consumption is at a minimum when the average external temperature is around 60°F. This also makes sense. When the weather is nice, the occupants shut down their heating.
- Then, the energy consumption also goes up when the average external temperature becomes higher. This is because the occupants start using their cooling.
- Finally, when the average external temperature is around 60°F, the energy consumption is at a minimum, but it does not become exactly zero.

Here is how you can turn these observations into mathematical statements:

- `hvac` ↓ if `t_out` < 60°F.
- `hvac` ↑ if `t_out` > 60°F.
- `hvac` is minimized when `t_out` is 60.

What is the simplest function that captures these trends? Well, that would be a parabola centered at 60°F. This one:

$$\text{hvac} = \text{hvac}(\text{t\_out}) = a(\text{t\_out} - 60)^2 + b,$$

for some parameters a and b that we need to pick.

This is our first model! And it is a function. Let's implement the function in Python[1]:

```python
def hvac_model(t_out, a, b, t_out_min=60):
    """A naive model of weekly HVAC energy consumption (kWh)
    as a function of external temperature t_out.
    The mathematical form of the model is:

        hvac = a * (t_out - t_out_min)^2 + b

    Arguments:
    t_out      -   The average external temperature in
                   degrees F (average over a week).
    a          -   A parameter to be calibrated using
                   observed data. In units of
                   kWh / (deg F)^2.
    b          -   Another parameter to be calibrated
                   using observed data. This is in
                   units of kWh. It is the energy
                   consumption when the HVAC system
                   is not used.
    t_out_min  -   The external temperature above
                   which the occupants feel
                   comfortable without using their
                   HVAC system.

    Returns: The weekly HVAC energy consumption in kWh.
    """
    return a * (t_out - t_out_min) ** 2 + b
```

5.4.1 *Manual model calibration*

The process of picking the parameters of a model by looking at the data is called *model calibration*. In machine learning literature, it is also called *training*.

---

[1]I am overdoing it with this huge docstring for such a simple model. But this is good practice. How else am I supposed to remember what the various inputs mean and why I created that function? Think about it.

You shouldn't be surprised that in engineering, the process is often done manually. First, engineers have a lot of domain knowledge and experience about the system they are modeling. They can use this knowledge to restrict the range of possible values for the parameters. Second, if the number of parameters is not too large, they can try different values and see which one fits the data better. The latter is not by any means the best way to do it, but it is quick, easy, and can help them develop intuition about the model. Let's give this manual calibration a try.

We have already decided to pick t_out_min = 60 degrees F because we see that the energy consumption is minimal at this temperature. We need to pick the parameters a and b. The parameter b is the energy consumption at the minimum. We can see also from the data that it should be around 50 kWh. So, we pick $b = 50$.

What about a? We know that the parabola should be looking like a cup. So, we need to pick a positive value for a. Trying a few values, and plotting the model against the data, we can find a value for a that visually fits the data. The value $a = 0.1$ seems to work well. The model is then:

$$\text{hvac} = 0.1 \times (\text{t\_out} - 60)^2 + 50.$$

See Figure 5.5.

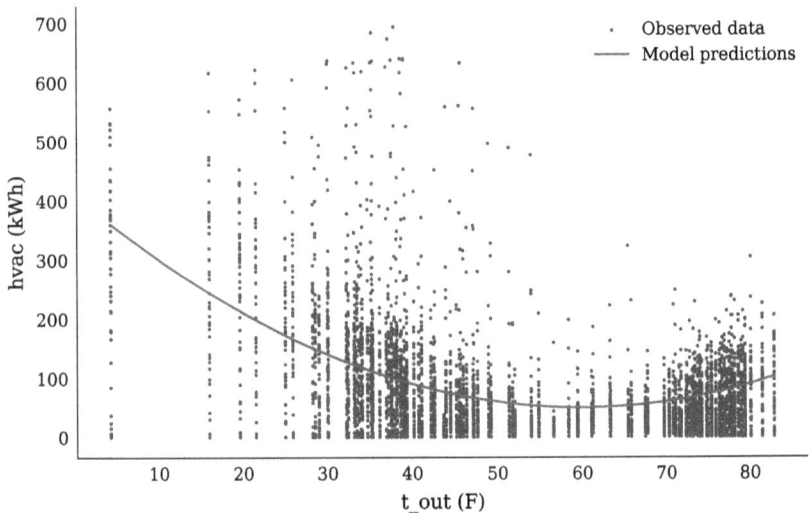

Figure 5.5.   The model `hvac_model(t_out, 0.1, 50)` plotted against the data.

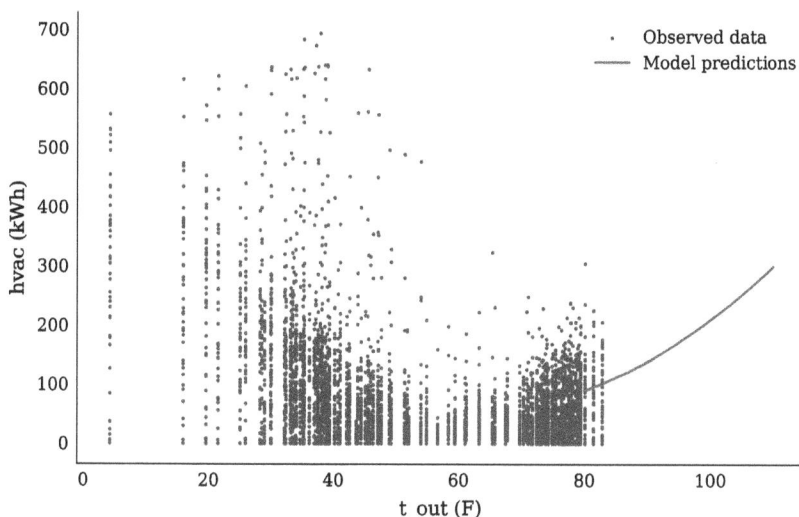

Figure 5.6. The model predictions for the weekly average HVAC energy consumption.

5.4.2 *Model predictions*

When we try to use the model with inputs that are outside the range of the data we used to calibrate it, we are making *predictions*. Let's make some predictions with the model we just built. Let's look at high temperatures which are missing form our dataset. Say from 80°F to 110°F, what weekly average HVAC energy consumption should we expect? We can make the predictions like this:

```
t_outs = np.linspace(80, 110, 100)
predicted_hvac = hvac_model(t_outs, a, b)
```

See Figure 5.6. Note that even though the prediction is unlikely to be correct, it can still be very useful. For example, you can use it to plan how much electricity load you are likely to need in an extreme heat event.

Exercises

1. Given variables `material` = "Steel" and `tensile_strength` = 400, create an f-string that outputs: "The tensile strength of Steel is 400 MPa."

2. Create an f-string that displays the part number 42 with a width of 5 characters, padded with zeros on the left (e.g., "00042").
3. Given a variable `efficiency` = 0.8756, create an f-string that formats it as a percentage with 2 decimal places: "87.56%."
4. Create an f-string that displays the result of calculating the area of a circle with radius 2.5 cm, showing 3 decimal places. Include units.
5. Given variables `voltage` = 120.45 and `uncertainty` = 0.32, create an f-string that displays the measurement with its uncertainty as: "120.45 ± 0.32 V." Make sure to format both values to 2 decimal places.
6. You are given a list of temperature measurements in Celsius. Write code to:

   (a) Display a header with the title "TEMPERATURE DATA" and a separator line.
   (b) Print each temperature value in scientific notation with 1 decimal place.
   (c) Calculate and display the average temperature with 2 decimal places.
   (d) Calculate and display the minimum and maximum temperatures with 2 decimal places.
   (e) Remove the docstring from the definition of `print_hello`, rerun the code block that defines it, and then try calling `help(print_hello)` again. Do you like the help you see now? That's why you have to have a docstring.
   (f) Modify the `print_hello_to` function to say hello to three people.
   (g) Modify the `plot_func` function so that you can also change the color with which it plots a function.
   (h) Write code that changes the `hvac` column of the `clean_data` dataframe from kWh's to MJ (megaJoule).

   Use the following data:

   ```
   temperatures = [23.0, 21.0, 23.0, 23.0, 23.84, 24.52]
   ```

   The expected output is:

   ```
   TEMPERATURE DATA
   ---------------
   2.3e+1
   ```

```
2.1e+1
...

Average: ...
Minimum: ...
Maximum: ...
```

7. **Analysis of experimental stress-strain curves of aluminum 6061-T651.** You are going to analyze the dataset collected by (Aakash *et al.*, 2019). The authors perform two types of experiments. We are going to focus on the "uniaxial tension experiment." This is what it is all about:

- They took several specimens of aluminum 6061-T651.[2]
- They mounted the specimen on a machine that applies tension.
- They controlled the temperature of the specimen.
- They applied tension gradually until the specimen broke, recording at each step the strain (% change in length) and the stress (force per cross section area in MPa, i.e., megaPascal).

I suggest that you skim through the paper if you want to understand more about the details of the experiment. Measuring the strains and stresses is not as straightforward as it sounds. Your goal is to download the data and, for a fixed temperature, create a model for the stress-strain relation. We are going to do some of the low level stuff. But, I am going to guide you through this.

(a) First, download the complete zipped data from the data repository of the paper and unzip it in the directory of your Jupyter notebook.
(b) Load the data into a pandas dataframe called `df`.
(c) Plot the stress as a function of the strain. Please label your axes properly.
(d) The ultimate tensile strength (or just "the strength") of a material is the maximum stress that develops under tension before the material breaks. What is the strength of this aluminum alloy? Please, provide your answer using the "print()" function with a precision of 2 decimal points. How does your

---

[2]A high-temperature aluminum alloy.

answer compare with the strength range for aluminum alloys reported in Wikipedia?

(e) Zoom in to low strains. Plot the first 200 observation points of the stress-strain curve.

(f) Observe that the experimental data are behaving strangely at very small strains. As a matter of fact, we are getting a few negative strains at the beginning. Let's throw these observations away. Start by finding the index i for which df['Strain'][i] becomes positive for the first time. You can do this by a visual inspection of df['Strain'][:30].

(g) Make a new dataframe, call it clean_df, where you have thrown away the initial data. Then plot the first 200 observations of clean_df.

(h) Observe that initially the stress-strain relation is linear. This is the so-called *elastic regime*. If you deform the material within this regime, it will return to its undeformed state without any permanent deformation effects. If you deform the material beyond the elastic regime, then you start having what we call *plasticity*. We are now going to focus exclusively on the linear regime. Find an index j so that clean_df[:j] is fully within the linear regime. Make a new dataframe, say linear_df containing only these data. You should probably pick j by visual inspection. And it is not important to pick the maximum j with this property. Just find one.

(i) In the elastic regime, the stress σ is a linear function of the strain ϵ:

$$\sigma = E\epsilon.$$

The constant E is called Young's modulus and it has units of GPa (GigaPascal). Make a function sigma(epsilon, E) that calculates the stress given the strain for any Young's modulus. Properly document the docstring of your function.

(j) If you pay close attention to the data in linear_df they do not cross zero. This is due to a systematic bias in the experiment. However, this bias is not important for calibrating Young's modulus E. Only the slope of the curve is important for finding E. So, let's make yet another dataframe (I promise you this is the last one) called clean_linear_df which removes this systematic bias. The dataframe clean_linear_df should contain the same data as linear_df but

- The strains should all be shifted by the minimum strain in `linear_df`. In other words, subtract from the strains in `clean_linear_df` the smallest strain in `linear_df`, i.e., subtract `linear_df['Strain'].min()`.
- The stresses should all be shifted by the minimum stress in `linear_df`.

(k) Use visual inspection to find a value for the Young's modulus that matches the data in `clean_linear_df`. Remember that the stress σ is in MPa so you will have to change the units correctly if you want E to be in GPa.

(l) How does what you find compare to the Wikipedia reported Young's modulus for this material? Why do you think yours is lower/higher? You may want to repeat the analysis above for another experiment with lower temperature (say at room temperature).

Chapter 6

Conditionals and Loops

In this chapter, I introduce logical expressions and demonstrate how to use Boolean expressions to extract dataframe rows that satisfy specific criteria. I also cover essential programming control structures: for-loops for iterating through sequences, while-loops for conditional iteration, and if-statements for controlling program flow based on conditions.

6.1   Python Conditionals: Colab

Conditional statements give you a way to change the flow of a Python program depending on the result of a *Boolean expression*, which is an expression that evaluates to either `True` or `False`.

6.1.1   *Boolean expressions*

Let's first see what are Boolean expressions. The simplest ones are:

```
>>> True
True
>>> False
False
```

You can also make them out of numerical comparisons:

```
>>> 1 > 2
False
```

```
>>> 25.0 <= 25.1
True
>>> 5 == 5
True
>>> 5 == 6
False
```

and so on.

You can also use variables in Boolean expressions:

```
>>> x = 'This is a string.'
>>> len(x) > 10
True
```

You can compare strings to each other:

```
>>> x == 'This is a string.'
True
>>> x == 'This is another string.'
False
```

The less-than and greater-than operators work as well:

```
>>> 'This' < 'This is a'
True
```

The comparison is done lexicographically.

So, what can you put in a Boolean expression? Well, anything that can be evaluated to either `True` or `False`. This is quite general, but very often we use the following building components:

- `>`: greater than.
- `>=`: greater than or equal to.
- `==`: equal to.
- `<`: less than.
- `<=`: less than or equal to.
- `not`: gives you the opposite of whatever Boolean expression follows.
- `or`: `True` if any of the Boolean expressions to the left or to the right are `True`.

- and: `True` if both of the Boolean expressions to the left and to the right are `True`.

You absolutely have to remember all these.

6.1.2 The simplest *if* statement

Now, let's look at the simplest conditional statement. It's syntax is as follows:

```
if <boolean_expression>:
    # Expressions that run if the bolean_expression is True
```

That's it. Let's try it:

```
x = 'Sort string'
if len(x) <= 20:
    print('The string x has less than 20 characters.')
```

If you run this, you will get:

```
The string x has less than 20 characters.
```

because the Boolean expression `len(x) <= 20` is `True`. The following does not print anything for the default 'x' because the Boolean expression is False:

```
if not len(x) <= 20:
    print('The string x does not have less than 20 characters.')
```

6.1.3 Python indentation

In Python, the empty spaces below an `if` statement are very important when you have multiple expressions. Here is an example that works:

```
if len(x) <= 20:
    print('The string x has less than 20 characters.')
    print('And this is an additional line.')
```

Notice how the second line is *indented* in exactly the same way as the first line. Here I use four spaces to indent the second line. The number of spaces does not matter, as long as it is the same for all the lines. Alternatively, you can use a tab character. But, if you do not use the exact indentation, you will get an `IndentationError`:

```
if len(x) <= 20:
    print('The string x has less than 20 characters.')
  print('And this is an additional line.')
```

6.1.3.1 *The `if-else` statement*

Some times you want to test for a Boolean expression and run something else if it is false. You can do this with the `if-else` statement:

```
if <boolean_expression>:
    # Expressions that run if the bolean_expression is True
else:
    # Expressions to run if the boolean_expression is False
```

Let's try it:

```
x = 'Sort string'
if len(x) <= 20:
    print('The string x has less than 20 characters.')
else:
    print('The string x does not have less than 20 characters.')
```

If you run this, you will get:

```
The string x has less than 20 characters.
```

6.1.4 *The `if-elif-else` statement*

Sometimes you want to test for multiple Boolean expressions. You can do this using the `if-elif-else` statement:

```
if <boolean_expression>:
    # Expressions that run if the bolean_expression is True
elif <other_boolean_expression>:
    # Expressions that run if the other_boolean_expression is True
    # if boolean_expression is False
else:
    # Expressions n to run otherwise
```

An example:

```
x = 'Sort string'
if len(x) <= 20:
    print('The string x has less than 20 characters.')
elif len(x) <= 30:
    print('The string x has between than 21 and 30 characters.')
else:
    print('The string x has more than 30 characters.')
```

If you run this, you will get:

```
The string x has less than 20 characters.
```

Note that you can have as many `elif` statements as you want and that `else` is always optional.

6.2 Python Loops: Colab

Loops are used when you need to run the same expression repeatedly for similar data. There are two kinds of loops in Python: `for` loops and `while` loops.

6.2.1 *The* `for` *loop*

The syntax of `for` loops is as follows:

```
for <variable> in <collection>:
    # run an expression potentially using <variable>
    # run another expression ...
```

The idea is that `<collection>` is an iterable object like a `tuple` or a `list` and that `<variable>` takes sequentially each one of the values in that iterable object. Here are a few examples:

```
for x in (1, 5, 6, 10, 20):
    print(f'x = {x:d}')
```

Output:

```
x = 1
x = 5
```

```
x = 6
x = 10
x = 20
```

Another one:

```
for c in 'Hello':
    print(f'c = {c}')
```

Output:

```
c = H
c = e
c = l
c = l
c = o
```

Very often, we want to loop over all values from 0 to a given number. We can do this with `range`:

```
for i in range(10):
    print(f'i = {i:d}')
```

Output:

```
i = 0
i = 1
...
i = 9
```

You don't have to start at 0:

```
for i in range(2, 10):
    print(f'i = {i:d}')
```

Output:

```
i = 2
i = 3
...
i = 9
```

And you can skip numbers:

```
for i in range(2, 10, 2):
    print(f'i = {i:d}')
```

Output:

```
i = 2
i = 4
i = 6
i = 8
```

And you can also go backwards:

```
for i in range(10, 0, -1):
    print(f'i = {i:d}')
```

Output:

```
i = 10
i = 9
...
i = 1
```

Lets now create some data and write code to calculate the average using a `for` loop. The data are:

```
data = [15.0, 0.25, 3.5, 5.1, 21.16]
```

The *average* of N numbers $x_1, x_2, \ldots, x_N$ is:

$$\bar{x} = \frac{x_1 + \cdots + x_N}{N}.$$

Here is how we can code this with a for loop:

```
N = len(data)
s = 0.0
for i in range(len(data)):
    x = data[i]
    s += x
average = s / N
print(f'Average = {average:1.2f}')
```

Output:

```
Average = 9.0
```

We have to explain a few things here. The variable `N` is the number of observations, which can just extract from the length of the data list. The variable `s` is used to accumulate the sum of all the elements of the data list as we go through them. Pay attention to the line `s += x`. This is just a shortcut for `s = s + x`.

It is always a good idea to compare something new you learned to an alternative implementation. Here is how we could have done the same thing using the built-in function `sum`:

```
average = sum(data) / len(data)
print(f'Average = {average:1.2f}')
```

If you run this code, you will get the same result as before. Do it.

Finally, here is another way to do the same calculation. Instead of using an index, we can directly go over each element in the list like this:

```
s = 0.0
for x in data:
    s += x
average = s / len(data)
print(f'Average = {average:1.2f}')
```

If you run this, you will also get the same result. Do it as well.

6.2.2   *The* `while` *loop*

The `while` loop is a bit more general than the `for` loop as stopping depends on a Boolean expression. Its syntax is like this:

```
while <boolean_expression>:
    # Some expressions to execute
    # Some other expression to execute
```

Let's see some examples of this:

```
i = 0
while i < 10:
    print(f'i = {i:d}')
    i += 1
```

Can you guess what this will print?

Let's do a more mathematical example. We are going to write code that calculates the sum of a series with a given tolerance. That is, given a sequence of real numbers: $a_1, a_2, \ldots$, we will write code that approximates:

$$a = \sum_{n=0}^{\infty} a_n.$$

To have more fun, we are going to sum a specific series:

$$1 - \frac{1}{3} + \frac{1}{5} - \frac{1}{7} + \cdots = \frac{\pi}{4},$$

and use it to approximate π.[1] In terms of the a_n notation above, the formula is:

$$\sum_{n=0}^{\infty} a_n = \sum_{n=0}^{\infty} \frac{(-1)^n}{2n + 1}.$$

Here is our first attempt:

```
max_iter = 100000
n = 0
a = 0.0
while n <= max_iter:
    a_n = (-1) ** n / (2 * n + 1)
    a += a_n
    n += 1
print(f'pi ~= {4 * a:1.12f}')
```

Let's break down what is happening here. The variable `max_iter` is the maximum number of iterations, which we set to 100,000. Whenever we are implementing a `while` loop, we need to make sure that there is a cap on the number of iterations. If you are not careful with this, you can easily create an *infinite loop*. The variable `n` is the current iteration index. Notice that the `while` loop will stop when `n` is greater than `max_iter`. This guarantees that the loop will terminate. The variable `a` accumulates the sum of the series. At each iteration,

---

[1]This is known as the Leibniz formula.

we calculate the next term in the series, a_n, add it to a, and increase n
by 1. If you run this, you will get a decent approximation of π:

```
pi ~= 3.141602653490
```

Whenever you are running loops that take a long time to run,
it is a good idea to print the current iteration index. Let's now
make our code print something every few iterations (say every 10,000
iterations).

```
max_iter = 100000
n = 0
a = 0.0
while n <= max_iter:
    a_n = (-1) ** n / (2 * n + 1)
    a += a_n
    if n % 10000 == 0:
        print(f'Current iteration n = {n:10d}, pi ~= {4 * a:1.12f}')
    n += 1
print(f'pi ~= {4 * a:1.12f}')
```

Notice that the only difference is that we added an if statement
inside the loop. This if statement checks if n is a multiple of 10,000.
It does it by calculating the remainder of the division of n by 10,000
and checking if it is 0. If it is, the code prints the current iteration
index and the current approximation of π. The output of this code
is:

```
Current iteration n =          0, pi ~= 4.000000000000
Current iteration n =      10000, pi ~= 3.141692643591
Current iteration n =      20000, pi ~= 3.141642651090
Current iteration n =      30000, pi ~= 3.141625985812
Current iteration n =      40000, pi ~= 3.141617652965
Current iteration n =      50000, pi ~= 3.141612653190
Current iteration n =      60000, pi ~= 3.141609319979
Current iteration n =      70000, pi ~= 3.141606939100
Current iteration n =      80000, pi ~= 3.141605153434
Current iteration n =      90000, pi ~= 3.141603764577
Current iteration n =     100000, pi ~= 3.141602653490
pi ~= 3.141602653490
```

We are going to end with a final modification: a *convergence test*. We are going to modify the code so that it loops over n until

$$|a_{n+1}| < \epsilon,$$

where ϵ is a small positive number typically called a *tolerance*. We can test for this condition using an `if` statement. But how can we stop the loop when this condition is met? We can do it by using the `break` command. The `break` command exits the loop immediately.[2] Here we go:

```
max_iter = 10000000
epsilon = 1e-6
n = 0
a = 0.0
while n <= max_iter:
    a_n = (-1) ** n / (2 * n + 1)
    if abs(a_n) < epsilon:
        print(f'*** Converged in {n+1:d} iterations! ***')
        break
    a += a_n
    if n % 100000 == 0:
        print(f'Current iteration n = {n:10d}, pi ~= {4 * a:1.12f}')
    n += 1
if n == max_iter + 1:
    print(f'*** Stopped when maximum number of iterations
    ↪ ({max_iter:d}) were reached! ***')
print(f'pi ~= {4 * a:1.12f}')
```

If you run this code, you will get:

```
Current iteration n =          0, pi ~= 4.000000000000
Current iteration n =     100000, pi ~= 3.141602653490
Current iteration n =     200000, pi ~= 3.141597653565
Current iteration n =     300000, pi ~= 3.141595986912
Current iteration n =     400000, pi ~= 3.141595153583
*** Converged in 500001 iterations! ***
pi ~= 3.141590653590
```

That's a much better approximation of π!

---

[2]Note that it also works in `for` loops.

6.3   Selecting Dataframe Rows That Satisfy a Boolean Expression: Colab

We are now going to put to use what we learned about Python Boolean expressions to extract rows from a dataframe that satisfy certain criteria.

6.3.1   *Extract rows that satisfy single Boolean expression*

Let's do this by example. Let's load again the `temperature_raw.xlsx` dataset we played with in Section 5.4:

```python
import pandas as pd
df = pd.read_excel('temperature_raw.xlsx')
df = df.dropna(axis=0)
df.date = pd.to_datetime(df['date'], format='%Y-%m-%d')
df.head()
```

Recall that we made a simple model between the weekly average of the consumed energy in kWh `hvac` and the average weekly external temperature `t_out`. The model was:

$$\text{hvac} = 0.1 \times (\text{t\_out} - 60)^2 + 50.$$

See Figure 6.1 for a visualization of the model. Clearly, this model represents the average energy consumption of the households. Let's use to see how specific households perform. Here is how many households we have:

```python
df['household'].unique()
```

Output:

```
array(['a1', 'a10', 'a11', 'a12', 'a13', 'a16', 'a2', 'a3', 'a4',
↪    'a5',
    'a6', 'a7', 'a8', 'a9', 'b17', 'b18', 'b19', 'b20', 'b21', 'b22',
    'b23', 'b24', 'b25', 'b26', 'b28', 'b29', 'b30', 'b31', 'b33',
    'c34', 'c35', 'c36', 'c37', 'c38', 'c39', 'c40', 'c41', 'c42',
    'c43', 'c44', 'c45', 'c46', 'c47', 'c48', 'c49', 'c50', 'a15',
    'a14', 'b32', 'b27'], dtype=object)
```

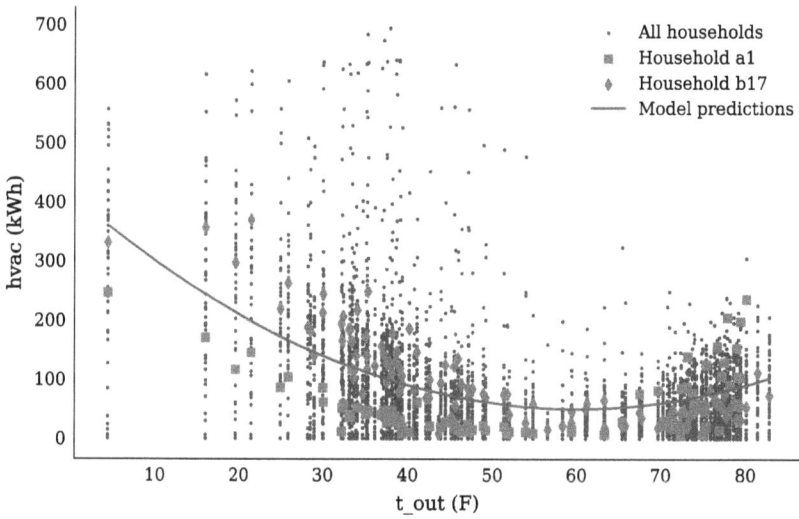

Figure 6.1. Scatter plot of the observed data and the model predictions for households **a1** and **b17**.

How can we extract the data for say household **a1**? You can do it as follows. First, notice that you can use a Boolean expression directly on the column **household**. The result is an array with `True` or `False` indicating the value of the Boolean expression on the corresponding rows. Here it is:

```
df['household'] == 'a1'
```

Output:

```
0        True
1        False
2        False
3        False
4        False
         ...
5643     False
5644     False
5646     False
5647     False
5649     False
Name: household, Length: 4887, dtype: bool
```

Now, if you feed this array of booleans to the dataframe, it will understand that you only want to keep the rows that are "True". Here it is:

```
df[df['household'] == 'a1'].round(2)
```

Output:

	household	date	score	t_out	t_unit	hvac
0	a1	2018-01-07	100.0	4.28	66.69	246.47
50	a1	2018-01-14	98.0	33.44	67.81	116.95
100	a1	2018-01-21	100.0	19.58	66.51	116.13
150	a1	2018-01-28	96.0	41.08	69.58	61.09
200	a1	2018-02-04	96.0	30.07	70.09	61.35
...	...	...	...	...	...	...
4700	a1	2019-10-27	97.0	54.83	73.62	8.56
4750	a1	2019-11-03	98.0	45.93	72.46	14.33
4800	a1	2019-11-10	97.0	40.18	72.45	10.76
4850	a1	2019-11-17	99.0	32.20	71.45	11.83
4900	a1	2019-11-24	97.0	39.33	70.90	8.46

82 rows × 6 columns

So, there are 82 rows corresponding to this household. If you wanted to find rows of another household, say b17, you do this:

```
df[df['household'] == 'b17'].round(2)
```

Output:

	household	date	score	t_out	t_unit	hvac
16	b17	2018-01-07	73.0	4.28	74.91	330.89
66	b17	2018-01-14	70.0	33.44	75.11	163.87
116	b17	2018-01-21	69.0	19.58	74.96	296.65
166	b17	2018-01-28	70.0	41.08	74.63	146.17
216	b17	2018-02-04	79.0	30.07	74.34	244.35
...	...	...	...	...	...	...
5416	b17	2020-02-02	92.0	33.66	70.94	101.67
5466	b17	2020-02-09	95.0	38.12	71.55	109.25
5516	b17	2020-02-16	92.0	28.55	71.43	182.11
5566	b17	2020-02-23	92.0	33.93	69.29	139.91
5616	b17	2020-02-25	95.0	43.64	72.25	8.29

113 rows × 6 columns

So, there are 113 rows corresponding to this household. The difference in the entries is likely due to sensor malfunction.

Now, let's redo our scatter plot but using different colors for `a1` and `b17`. We can do this by using the `hue` parameter in the `sns.scatterplot` function. Here is how we do it:

```
fig, ax = plt.subplots()
ax.scatter(df['t_out'], df['hvac'], label='All households', s=0.5)
a1_df = df[df['household'] == 'a1']
ax.scatter(a1_df['t_out'], a1_df['hvac'], label='Household a1',
↪   s=10, marker='s')
b17_df = df[df['household'] == 'b17']
ax.scatter(b17_df['t_out'], b17_df['hvac'], label='Household b17',
↪   s=10, marker='d')
ax.set_xlabel('t_out (F)')
ax.set_ylabel('hvac (kWh)')
predicted_hvac = hvac_model(t_outs, a, b)
ax.plot(t_outs, predicted_hvac, 'r', label='Model predictions')
plt.legend(loc='best')
```

Notice the difference in behavior between the two households. Household `a1` is using less energy during the heating season, but more during the cooling season.

6.3.2 *Plotting timeseries data*

Since we have extracted the data for units `a1` and `b17`, we have a good opportunity to demonstrate another useful plot that you can do with dataframes: plotting timeseries data. A *timeseries* is a sequence of data points indexed in time. This is exactly what we have here. For each unit, we can look at energy consumption as a function of time, or the unit temperature as a function of time, etc. By plotting data as a function of time, we can see temporal patterns. Let's plot the unit temperature for the two households (Figure 6.2). Here is how we do it:

```
fig, ax = plt.subplots()
a1_df.plot(x='date', y='t_unit', label='Household a1', style='.-',
↪   ax=ax)
```

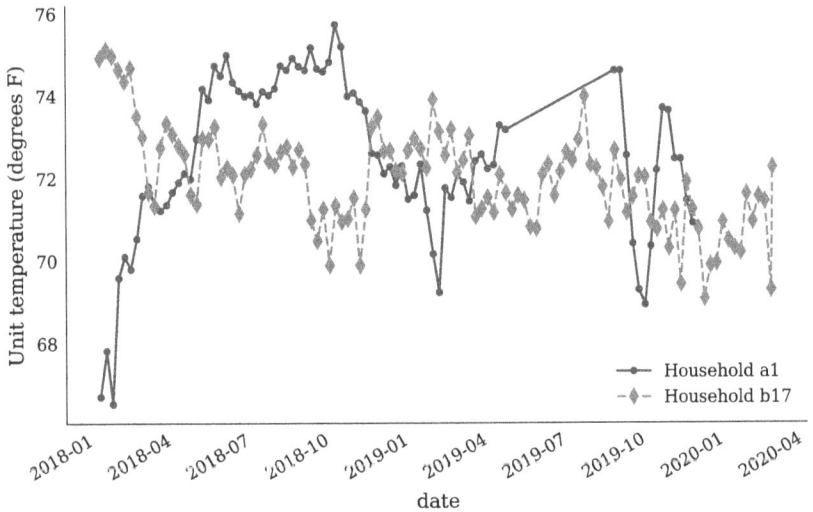

Figure 6.2. Timeseries plot of the unit temperature for households **a1** and **b17**.

```
b17_df.plot(x='date', y='t_unit', label='Household b17',
↪   style='d--', ax=ax)
ax.set_ylabel('Unit temperature (degrees F)')
```

First, observe that `pd.DataFrame.plot()` *linearly interpolates* the data. This means that it draws a straight line between each pair of points. This is why the plot looks a bit jagged. In the plots above, I have used markers to make the points more visible. Notice that for a1, there is a gap in the middle of the data, but the temperature is still linearly interpolated, and that the plot also stops earlier.

As you analyze real data, you will often find gaps like this. It is important to understand why these gaps exist before you move on to more complex analyses. Why is there a data loss? Was there a power or Wi-Fi outage? Are the sensors corrupted? Do you have to go back and collect better data? Or can you ignore these gaps and proceed? If your data are wrong, everything that follows is useless. In this case, we can ignore the gaps. The gap was actually due to Wi-Fi outage.

So far we manually extracted the data we needed for each unit. Now, I will show you how you can do it for as many units as you like using a `for` loop. Let's say we have a list of households we want to plot. We need to plot the data for each household using a different marker. How can we loop over pairs of household names and markers? The most elegant way to do this is by using the `zip` function:

```
households = ['a1', 'a2', 'a3', 'a4', 'a5']
markers = ['o', 's', 'd', 'p', 'x']
for h, m in zip(households, markers):
    print(h, m)
```

Output:

```
a1 o
a2 s
a3 d
a4 p
a5 x
```

The `zip` function takes two lists and returns a list of pairs.[3] Each pair contains one element from the first list and one from the second. Let's redo our scatter plot for a list of households (Figure 6.2):

```
fig, ax = plt.subplots()
ax.scatter(df['t_out'], df['hvac'], label='All households', s=0.5)
households = ['a1', 'a2', 'a3', 'a4', 'a5']
markers = ['o', 's', 'd', 'p', 'x']
for h, m in zip(households, markers):
    df_tmp = df[df['household'] == h]
    ax.scatter(df_tmp['t_out'], df_tmp['hvac'], label='Household ' +
    ↪  h, marker=m)
ax.set_xlabel('t_out (F)')
ax.set_ylabel('hvac (kWh)')
plt.legend(loc='best')
```

6.3.3 *Extract rows that satisfy more complicated Boolean expressions*

Let's say that we want to see how the selection of `t_unit` (temperature setpoint) affects `hvac` (energy consumption) during the heating

---

[3]Actually, you get back a object called an *iterator*, but you can think of it as a list of pairs.

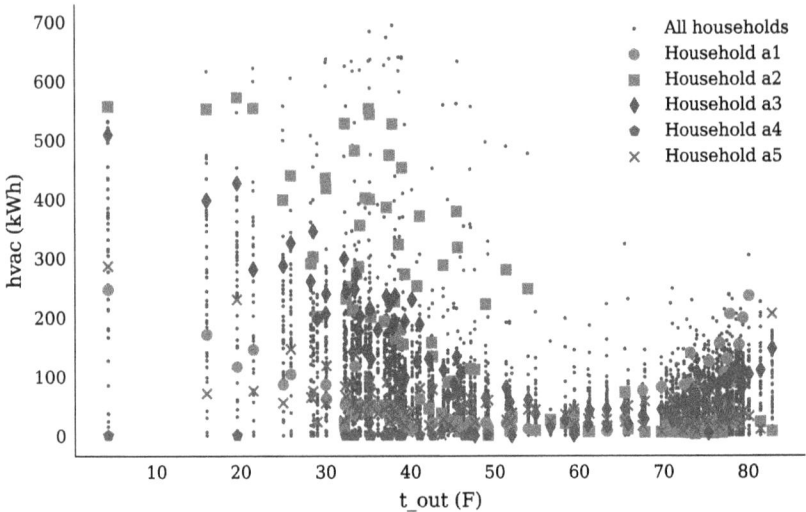

Figure 6.3. Scatter plot of the observed data and the model predictions for households a1, a2, a3, a4 and a5.

season. We are on heating season for sure when the external temperature is smaller than 55°F. So, we need to extract all the rows with df['t_out'] < 55. Let's do that:

```
df_heating = df[df['t_out'] < 55]
df_heating.round(2)
```

Output:

```
     household  date         score   t_out   t_unit   hvac
0         a1    2018-01-07   100.0    4.28    66.69    246.47
1        a10    2018-01-07   100.0    4.28    66.36      5.49
2        a11    2018-01-07    58.0    4.28    71.55    402.09
3        a12    2018-01-07    64.0    4.28    73.43    211.69
4        a13    2018-01-07   100.0    4.28    63.92      0.85
...      ...    ...          ...      ...     ...       ...
5643     c44    2020-02-25    59.0   43.64    76.49     19.14
5644     c45    2020-02-25    87.0   43.64    71.17     30.79
5646     c47    2020-02-25    97.0   43.64    68.60      5.34
5647     c48    2020-02-25    92.0   43.64    73.43     18.04
5649     c50    2020-02-25    59.0   43.64    77.72     14.41
     2741 rows x 6 columns
```

Now, we are going to draw the scatter plot between `t_out` and `hvac`, but coloring differently those households with `t_unit` within the range 70 to 72°F. First, let's extract the rows with `t_unit` between 70 and 72°F:

```
df_heating_70to72 = df_heating[(df_heating['t_unit'] >= 70) &
                               (df_heating['t_unit'] < 72)]
df_heating_70to72.head().round(2)
```

Output:

```
household  date      score  t_out  t_unit  hvac
2          a11       58.0   4.28   71.55   402.09
17         b18       87.0   4.28   70.61   368.15
25         b26       97.0   4.28   70.01   36.65
29         b30       99.0   4.28   71.42   23.68
34         c35       96.0   4.28   70.56   349.03
```

Pay attention to two things in this code. First, notice that we do not use the `and` operator, but the *bitwise "and"* operator, `&`. As a matter of fact, if you use the `and` operator, you will get an error. This is because the `and` operator works only with booleans and expressions like `df_heating['t_unit'] >= 70` return a `pandas.Series` of booleans. You can think of a series of booleans as an array of bits, where each bit is either 0 or 1 (`True` or `False`). The `&` operator works element-wise on bits. For example,

```
a = pd.Series([True, False, True])
b = pd.Series([True, True, False])
a & b
```

Output:

```
0    True
1    False
2    False
dtype: bool
```

Second, notice that I put each condition in parentheses. This is absolutely necessary because the `&` operator has higher precedence than the comparison operators. Consider the following (wrong) expression:

```
df_heating['t_unit'] >= 70 & df_heating['t_unit'] < 72
```

When Python sees this, it will attempt to evaluate 70 &
df_heating['t_unit'] first which makes no sense.[4]

Now I can do the scatter plot I was after:

```
fig, ax = plt.subplots()
df_heating_70to72.plot(x='date', y='hvac', label='Households with
↪   t_unit between 70 and 72 degrees F', style='.-', ax=ax)
ax.set_ylabel('hvac (kWh)')
plt.legend(loc='best')
```

See Figure 6.4 for the plot. We see that the units with t_unit between
70 and 72°F tend to consume less energy. There are some outliers, but
the trend is clear. What can explain the outliers though? How can
two households with the same internal temperature t_unit have very
different energy consumption? There are other variables here that
we are not observing. For example, the location of the unit in the
building. A unit that is at the corner of the building will have different

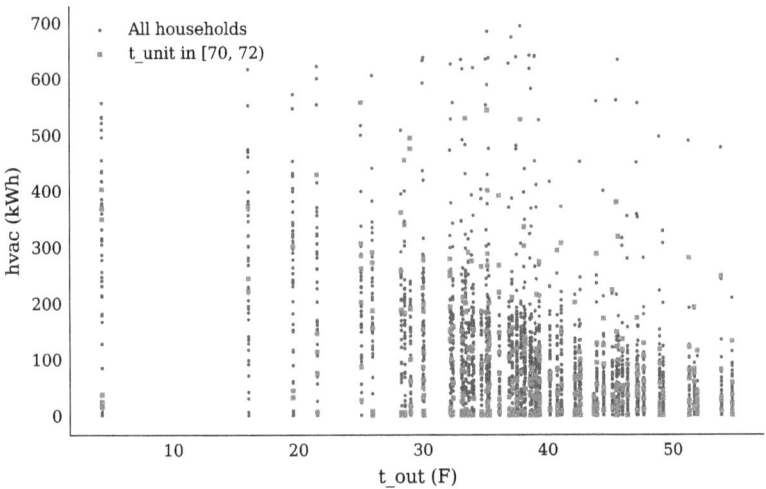

Figure 6.4. Scatter plot of the observed data and the model predictions for
households with t_unit between 70 and 72°F.

---

[4]Note that we do not have this problem with the and operator because it has
lower precedence than the comparison operators. That is, writing 3 < 6 and 5
< 10 is perfectly fine.

heat loss than a unit that is in the middle of the building. Another explanation is that the occupants may be opening the windows more often in one of the apartments. Such variables are called *confounding variables*. You should always be on the lookout for them.

Let's demonstrate a logical-or for selecting rows. As you may have guessed `or` does not work. We need to use the bitwise-or operator `|`. And yes, we need to enclose everything in parentheses. Let's select the entries that highlight wasteful energy consuming behavior. That would be units that have a very high `t_unit` (say above 78°F) during heating or a very low `t_unit` during cooling (say below 70°F). Let's see how much energy they consume:

```
df_bad_behavior = df[((df['t_out'] < 55) & (df['t_unit'] > 78)) |
                ((df['t_out'] >= 55) & (df['t_unit'] < 70))]
df_bad_behavior.head().round(2)
```

Output:

	household	date	score	t_out	t_unit	hvac
15	a9	2018-01-07	39.0	4.28	78.71	495.23
22	b23	2018-01-07	44.0	4.28	78.62	359.17
65	a9	2018-01-14	41.0	33.44	78.56	321.83
72	b23	2018-01-14	43.0	33.44	78.68	197.92
115	a9	2018-01-21	41.0	19.58	78.57	431.24

And the scatter plot:

```
fig, ax = plt.subplots()
ax.scatter(df['t_out'], df['hvac'], label='All households', s=0.5)
ax.scatter(df_bad_behavior['t_out'], df_bad_behavior['hvac'],
        label='Bad behavior', marker='s', s=2)
ax.set_xlabel('t_out (F)')
ax.set_ylabel('hvac (kWh)')
plt.legend(loc='best')
```

See Figure 6.5 for the plot. Notice that there are some entries that exhibit bad behavior without being penalized with excessive energy consumption. Can you think of a confounding variable that could explain this? Here is what is happening:

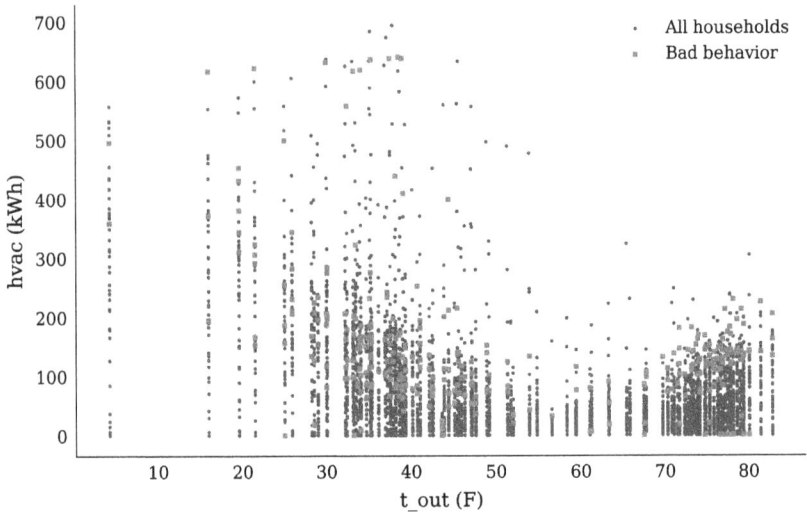

Figure 6.5. Scatter plot of the observed data and the model predictions for households with bad behavior.

- Units that are at the corners of the building expose at least two sides to the external environment.
- Units that are on the top floor expose at least two sides to the external environment.
- Units that are at a corner and on the top floor expose three sides to the external environment.
- The rest of the units expose a single side to the external environment.

The first three groups (and especially the third group) are loosing a lot of energy to the environment and their energy consumption will in general be higher. The final group is basically insulated from the environment and on top of that they may be getting heat energy from their neighbors. Behaving badly in a top floor, corner unit will be reflected on your energy bill. Behaving badly in an insulated unit may have no effect on your energy bill.

Let's end with logical-not for selecting rows. Say we wanted to negate the previous selection so that we highlight all those who behaved well. For this, you use the bitwise-not operator ⁻. Like this:

```
df_good_behavior = df[~(((df['t_out'] < 55) & (df['t_unit'] > 78)) |
                         ((df['t_out'] >= 55) & (df['t_unit'] <
                         ↪ 70)))]
df_good_behavior.head().round(2)
```

Output:

```
   household  date        score   t_out   t_unit   hvac
0  a1         2018-01-07  100.0   4.28    66.69    246.47
1  a10        2018-01-07  100.0   4.28    66.36    5.49
2  a11        2018-01-07  58.0    4.28    71.55    402.09
3  a12        2018-01-07  64.0    4.28    73.43    211.69
4  a13        2018-01-07  100.0   4.28    63.92    0.85
```

Again, pay attention to the parentheses. Make the corresponding scatter plot and experiment with the definition of what constitutes "good" and "bad" (energy consuming) behavior. The plot is shown in Figure 6.6. Most households behaving well consume less than 100 kWh of energy per week. However, there are some units that are using excellent temperature set points, but they are still consuming a lot of energy. This could be because of the confounding variables

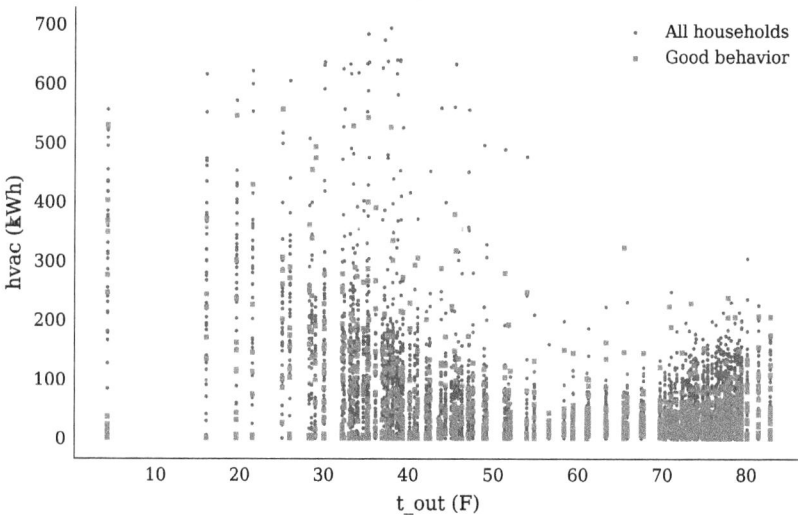

Figure 6.6. Scatter plot of the observed data and the model predictions for households with good behavior.

we discussed earlier. For example, the unit could be in a corner and
the occupants could be opening the windows more often.

Exercises

1. Write a for loop that calculates the average of all the elements in
   the list data that are greater than 5.
2. Write a for loop to calculate the *second moment* of the data list,
   i.e., this expression:

$$\bar{x}_2 = \frac{x_1^2 + \cdots + x_N^2}{N}$$

3. Write a for loop that finds the maximum of the data list.
4. Write a while loop that approximates sum the series:

$$\sum_{n=0}^{\infty} \frac{1}{2^n},$$

   which converges to 2.
5. Remake Figure 6.2 adding a new line for household c36. Make sure
   to use a different marker.
6. For households a1, b17 and c36 plot hvac as a function of the date.
7. Remake Figure 6.3 for all the b?? households.[5] Repeat for hvac.
8. Consider the following list:

   data = [1, 4, 3, 10, 4, 3, 4, 4]

   (a) Write a loop that computes the average of the elements in the
       list and print the result using two significant digits.
   (b) Write code that finds the number of times the element 4 occurs
       in the list. *Hint*: Use a loop and an if-statement.
   (c) Write a Python function that takes a list as an argument and
       returns the number of times a given element (also passed as an
       argument to the function) appears in the list. Call that func-
       tion find_number_of_occurences(a, elm). Make sure you follow

---

[5]The ?? is a wildcard that matches any two numbers. So you are supposed to
select all the households that start with b and have two more numbers, e.g., b17,
b23, etc.

best practices when writing the docstring of your function. Try your function with the following code:

```
n = find_number_of_occurences(data, 4)
print(f'The number of times 4 occurs in the list is {n}')
```

(d) Write a Python function that takes a list as an argument and returns the number of elements that are greater than a given element (also passed as an argument to the function). Call that function `find_number_of_elms_greater_than(a, elm)`. Make sure you follow best practices when writing the docstring of your function. Try your function with the following code:

```
n = find_number_of_elms_greater_than(data, 3)
print(f'The number of elements greater than 3 is {n}')
```

9. In this problem we will continue analyzing the high-performance buildings dataset, `temperature_raw.xlsx`.

(a) Plot the external temperature `t_out` as a function of time.

(b) Extract the data pertaining to household `a5`. Put the result in a new dataframe called `df_a5`.

(c) For household `a5`, plot `t_unit` as a function of date.

(d) In a single figure, plot date vs `t_unit` for households `a5` and `a11`.

(e) In the same figure, plot the `t_out` vs `hvac` scatter plots for both households `a5` and `a11`.

(f) In the same figure, plot the histogram of `t_unit` for households `a5` and `a11`. Which household prefers cooler temperatures? *Hint*: To make the histogram more appealing use the keywords `density=True` and `alpha=0.25`.[6]

(g) In the same figure, plot the histogram of `hvac` for households `a5` and `a11`. Which household is more energy efficient (if any) and why?

(h) Repeat the analysis above for households `b17` and `c40`. Which household prefers cooler temperatures and which one is more energy efficient?

---

[6]The keyword `alpha` sets the transparency of the histogram bars.

Chapter 7

Probability as a Measure of Uncertainty

> Probability theory is nothing
> but common sense reduced to
> calculation.

Pierre-Simon Laplace, *Théorie
analytique des probabilités*
(1814)

In this chapter, I explore probability as a representation of our state of knowledge, learning how to assign probabilities to simple events using the principle of insufficient reason. I explain how empirical frequencies can be used to estimate probabilities and demonstrate the convergence of these frequencies to true probabilities as sample sizes increase.

7.1 Probability as a Representation of Our State of Knowledge

Probability theory allows us to quantify our uncertainty about something. The "something" can be a lot of different things. It can be predicting tomorrow's weather, assessing cracks in a bevel gear, evaluating the structural integrity of an extraterrestrial habitat, analyzing the predictions of physical models, examining the validity of mathematical theories, or evaluating any statement about the world. Paraphrasing the title of (Jaynes, 2003), *probability theory is the logic of science.*

7.1.1 *Types of uncertainty*

We are uncertain about something if we don't know everything about it. This kind of uncertainty is called *epistemic uncertainty*. The term *epistemic* comes from the Greek word $\epsilon\pi\iota\sigma\tau\acute{\eta}\mu\eta$ (epistēmē), which means *knowledge*. Epistemic uncertainty is associated with lack of knowledge.

It is easy to find examples of epistemic uncertainty. Consider this simple demonstration: I take a coin in my hand, place my hands behind my back, and conceal the coin in one of my closed fists before presenting both hands to you. You don't know which hand contains the coin. Yet the coin is definitely in one of my hands. This uncertainty doesn't arise from randomness. It stems purely from your lack of information about where I placed the coin.

In the same spirit, consider a mechanical component like an aircraft engine turbine blade that may have developed microscopic cracks. The blade either has these cracks or it doesn't. There is no randomness in its actual condition. The uncertainty exists solely because you haven't yet performed the necessary inspection with specialized equipment to detect these cracks.

You can, in principle, reduce epistemic uncertainty if you are willing to pay the price, e.g., by collecting more data, building a new sensor, doing a better analysis of the data. In practice, this is not always possible. For example, the only way to know where exactly I placed the coin is to place some sensors on my hands or perhaps read my mind.

Uncertainty can also come from irreducible randomness. By irreducible randomness, I mean that the outcome is not determined perfectly by complete knowledge of the initial conditions. This kind of uncertainty is called *aleatory uncertainty*. The term aleatory (sometimes also *aleatoric*) comes from the Latin word *alea*, which means *dice*. People chose this term because throwing a dice seemed like a good example of something inherently random. But, it is not. Using principles of classical mechanics, you can describe the dynamics of dice and predict exactly what is going to happen in a dice throw experiment if you know the initial conditions.[1] The uncertainty about the result of a dice throw is not aleatory. It is epistemic uncertainty, i.e., lack of knowledge about what the initial conditions are.

You have to search hard for true aleatory uncertainty. Perhaps, the closest you can get is if you look at very small things. So small that quantum effects start being important. At this most fundamental level, the leading interpretation of quantum mechanics suggest that nature is inherently random.[2]

There is a long (and ongoing) philosophical debate about the distinction between aleatory and epistemic uncertainty. Is this or that event truly random or is it just that we don't know enough about it? We are going to ignore this debate because probability theory is sufficient to describe both types of uncertainty and that the distinction in many cases is not important.

7.1.2 *Probability as a measure of uncertainty*

Let's make the "something" on which probability theory is applied more precise. We will restrict it to *logical propositions*. A logical proposition is a statement about the world that is either true or false. For example, "The result of the next coin toss Yannis performs will be heads" is a logical proposition. So is "On 7/1/2021, the average temperature in West Lafayette, IN, was greater than 70°F". Or "The radiator of the cooling system has paint damage." The only requirement that we are going to impose on such logical propositions is that they are not weird. By weird, I mean propositions like "This sentence is false." In general, if you can think of a way to test the truth of a proposition by performing an experiment, then it is not weird.

What do I mean by "quantify our uncertainty?" I want a mathematical theory that can turn all the *information* I have into a real

---

[1] Persi Diaconis, a Stanford mathematician (and former professional magician), explains why coin tosses are not random in the YouTube video How random is a coin toss? - Numberphile. The same principles can be used to predict the outcome of a dice throw.

[2] Watch this YouTube video titled Double Slit Experiment explained!, by Jim Al-Khalili, a theoretical physicist from the University of Surrey in the UK.

number that tells me how plausible it is that the logical proposition is true.

What do I mean by "all the information we have?" I am talking about what I know about the world at the moment I am making the assessment. And I am talking about absolutely everything: what my parents taught me, what I learned in school, what I learned in college, any data I may have seen, etc.

I am about to turn everything into math. I will use the letter I to denote all the information we have and A to denote a logical proposition. Probability theory will give us a number $p(A|I)$, which we read as:

$p(A|I)$ = the probability that A is true given that we know I.

You can think of this number as the degree of belief that we have in the proposition A given the information I. Pay attention to the | symbol. It is read as "given that."

7.2   Probability from Common Sense

How do you come up with the number $p(A|I)$? And how do you combine it with other probabilities to make a coherent argument? (Jaynes, 2003) goes through a thought experiment to answer these questions.

Suppose you want to make an intelligent agent that can argue under uncertainty. It should be able to take logical propositions and argue about them using all the information you have. What guidelines should you give to this agent? Jaynes suggests this agent should have the following properties:

(1) The agent should represent its uncertainty about a proposition using a real number.
(2) The agent should have a qualitative correspondence to common sense.
(3) The agent should be consistence in the sense that:
  (i) If a conclusion can be reached in two ways, each way must leads to the same result.
  (ii) All evidence relevant to a question should be taken into account.

(iii) Equivalent states of knowledge must be represented by equivalent plausibility assignments.

The first property simply says that $p(A|I)$ should be a real number. This is not a unique choice. One can envision other possibilities. For example, you can use a complex number to represent the degree of belief in a proposition. This is what is done in quantum mechanics. But using a real number seems like the simplest choice.

The second property is more subtle. It says that the agent should have a qualitative correspondence to common sense. What does Jaynes mean by that? Let me give you some examples of common sense. Suppose you have another logical proposition B that is logically equivalent to A. Then, you should have the same degree of belief in A and B. Or:

$$(A \equiv B) \implies (p(A|I) = p(B|I)).$$

For example, take A to be "it will rain tomorrow" and B to be "water will fall from the sky tomorrow". You should have the same degree of belief in both propositions. They are both saying the same thing.

What is another example of common sense? Suppose now that B implies A. If you know that B is true, then you should believe that A is true. Or:

$$(B \implies A) \implies (p(A|BI) = 1).$$

Here with $p(A|BI)$ I mean the probability of A being true given that B is true and I is true. We say that we *condition* on B and I. For example, take B to be "it is midday and there are no clouds in the sky" which implies A "the sun is shining." So, if you know that it is midday and there are no clouds in the sky, you increase your belief in the sun shining.[3]

The third property is self-explanatory. Read it again and pause to think about it.

Jaynes shows that these properties are enough to derive the mathematical foundations of probability theory. The theory assigns numbers to mathematical expressions like $p(A|I)$, $p(B|I)$, $p(A|BI)$, etc.

---

[3]You need to also know that there is no solar eclipse happening.

Note that the mathematical theory is not unique. There is some indeterminacy equivalent to picking units.

I will follow the common assumption of using zero for impossibility and one for certainty. So, the probability $p(A|BI)$ is a number between zero and one quantifying the degree of plausibility that A is true given B and I. Specifically:

- $p(A|B, I) = 1$ when we are certain that A is true if B and I are true.
- $p(A|B, I) = 0$ when we are certain that A is false if B and I are true.
- $0 < p(A|B, I) < 1$ when we are uncertain about A if B and I are true.
- $p(A|B, I) = \frac{1}{2}$ when we are completely ignorant about A if B is true (and I).

7.3   The Principle of Insufficient Reason

The *principle of insufficient reason*[4] tells us how we should pick probabilities for simple logical propositions before we see any data. It is often the starting point of any analysis. The idea is straightforward. Take I to be the background information and A be a logical proposition. Then the principle states:

$$p(A|I) = \frac{\text{Number of ways } A \text{ can be true under } I}{\text{Number of things that can happen under } I}.$$

The idea is best illustrated through examples.

7.3.1   *Coin toss experiment*

Let's look first at a simple coin toss experiment. The background information I you typically have is like this:

> Bob has a coin in his pocket. I have no reason to believe that it is biased. Bob puts his hand in his pocket, shakes it for a while, and then makes a fist around the coin. He takes his hand out

---

[4]The idea dates back to Pierre-Simon Laplace (1749–1827), a French mathematician and astronomer and one of the founders of the theory of probability.

with the coin hidden in his fist. He then throws the coin up and
he lets it fall on the ground and stop moving.

Now take the logical proposition H:

The top face of the coin shows heads.

The principle of insufficient reason tells you what number you should
assign to the probability of heads given the information you have.
Since there are two possible results and heads is one of them, the
principle dictates that you should assign probability $1/2$, i.e.,

$$p(H|I) = \frac{1}{2}.$$

7.3.2 Six-sided die

The background information I is:

Alice has a six-sided die in her pocket. The die is well-balanced.
She puts her hand in her pocket, shakes it for a while, and then
makes a fist around the die. She takes her hand out with the
die in her fist. She throws the die up and she lets it fall on the
ground and stop moving.

Take the logical proposition A:

The top face of the dice shows the number 1.

Since A is one of the six events allowed under I, then by the principle
of insufficient reason you should pick:

$$p(A|I) = \frac{1}{6}.$$

7.3.3 Two six-sided dice

Let's make things a bit more complicated by taking two dice. The
background information I is:

Alice has two six-sided dice in her pocket. The dice are well-
balanced and indistinguishable. She puts her hand in her
pocket, shakes them for a while, and then makes a fist around
the dice. She takes her hand out with the dice in her fist. She
throws the dice up and she lets them fall on the ground and
stop moving.

Now take the logical proposition B:

> The top face of one die shows the number 1 and the top face of
> the other die shows the number 2.

In how many ways can this happen? There are 36 possible outcomes
$(36 = 6 \times 6)$ and out of those, the logical proposition B is true two
times (either the first die shows 1 and the second 2 or the first shows
2 and the second 1). So:

$$p(B|I) = \frac{2}{36} = \frac{1}{18}.$$

Let's do one more. Consider the logical proposition C:

> The sum of the top faces of the dice equals 7.

Again, there are 36 possibilities under I. We can get a total sum of
7 in the following 6 ways:

- Die 1: 1, Die 2: 6
- Die 1: 2, Die 2: 5
- Die 1: 3, Die 2: 4
- Die 1: 4, Die 2: 3
- Die 1: 5, Die 2: 2
- Die 1: 6, Die 2: 1

So, by the principle of insufficient reason, we can assign:

$$p(B|I) = \frac{6}{36} = \frac{1}{6}.$$

7.4   Estimating Probabilities by Simulation: Colab

I will teach you now a very powerful method for estimating probabilities that is typically reserved for much later in probability courses.
Proving why it works requires advanced mathematics. But develop-
ing intuition about the method is trivial. Also, implementing it just
requires that you understand loops and conditionals. So let's not
wait! Let's learn about Monte Carlo simulations!

Say that you want to calculate the probability $p(A|I)$ of a logical
proposition A given some background information I. The idea is to
write computer code that simulates the experiment of measuring A

under I, run it many times, and calculate the frequency with which A is true under I. So, you set:

$$p(A|I) \approx \frac{\text{Number of times } A \text{ is True under } I \text{ in } N \text{ simulated experiments}}{N}.$$

Now the larger the number of simulated experiments N, the closer your estimate would be to the true probability $p(A|I)$. This is a consequence of a theorem called the *law of large numbers*. You can learn about it in a probability theory course. Let's see this working in practice.

7.4.1 *Simulating a coin toss experiment*

I am going to simulate a coin toss experiment. I need the ability to randomly pick between the numbers 0 and 1. This is sufficient because I can associate 0 with tails and 1 with heads. Python has a function that can generate random integers for me. It is called numpy.random.randint. Let's see how it works before I set up my coin toss experiment.

The simplest way to call `randint()` is `randint(low, high)` where `low` and `high` are two integers. Then it will respond with a random integer between `low` and `high` - `1`. So, if I want to pick randomly between 0 and 1, I have to call it like this:

```
import numpy as np
np.random.randint(0, 2)
```

If you want to call it multiple times, you can either make a loop:

```
for i in range(10):
    print(np.random.randint(0, 2))
```

with the output:

```
0
1
0
1
1
1
0
```

```
0
1
1
```

Or you can call it by specifying the keyword argument `size`:

```
np.random.randint(0, 2, size=10)
```

Output:

```
array([0, 1, 0, 0, 0, 1, 0, 1, 0, 1])
```

Now, you may wonder: "Computers are deterministic machines. How can they generate random numbers?" Well, they can't... These are not true random numbers. They are pseudo-random numbers! But they look like random numbers, so they are good enough for simulating experiments. What you need to know is that there is a thing called ꜱeed that uniquely specifies the sequence of random numbers you get. The seed can be any integer. Look at this:

```
np.random.seed(12345)
np.random.randint(0, 2, size=20)
```

The output is:

```
array([0, 1, 1, 1, 0, 1, 0, 0, 1, 0, 1, 1, 0, 1, 1, 0, 1, 1, 1, 0])
```

If I run the second line again, you will get a different output:

```
np.random.randint(0, 2, size=20)
```

Output:

```
array([0, 0, 1, 1, 1, 0, 1, 0, 0, 1, 1, 1, 1, 1, 0, 1, 1, 1, 1, 1])
```

But if I fix the seed again (to the same integer 12345), I get exactly the same results:

```
np.random.seed(12345)
np.random.randint(0, 2, size=20)
```

Output:

```
array([0, 1, 1, 1, 0, 1, 0, 0, 1, 0, 1, 1, 0, 1, 1, 0, 1, 1, 1, 0])
```

Remember: when writing code that simulates random experiments it is always a good idea to fix the random seed at the very beginning. This ensures that other people who run your code will get exactly the same results as you do!

Okay, let's go back to the coin toss experiment. How do you simulate it. You just call `np.random.randint(0, 2, size=N)` for any N you like and that's it!

Let's now use this to estimate the probability of heads. To get the estimate we have to divide the number of heads with N. Since we are associating heads with the experimental result 1, then we can find the number of heads by summing up the result of all the experimental runs. So, we are doing this:

$$p(H|I) \approx \frac{\text{result } 1 + \ldots \text{result } N}{N}.$$

This is easy. Here we go:

```
# Set the seed for reproducibility
np.random.seed(12345)
# Pick the number of coin toss experiments you want to simulate
N = 5
# Simulate the experiments
coin_toss_exp = np.random.randint(0, 2, size=N)
# Approximate the probability
p_H_g_I = coin_toss_exp.sum() / N
# Print the estimate
print(f'p(H|I) ~= {p_H_g_I:1.5f}')
```

Output:

```
p(H|I) ~= 0.60000
```

This seems quite off from what we expected (0.5). But, hey, we just used $N = 5$ experiment. We need to use much more than this. Let's repeat with $N = 1000$:

```
np.random.seed(12345)
N = 1000
coin_toss_exp = np.random.randint(0, 2, size=N)
p_H_g_I = coin_toss_exp.sum() / N
print(f'p(H|I) ~= {p_H_g_I:1.5f}')
```

```
p(H|I) ~= 0.50800
```

Okay, this is almost perfect.

How do you know how many experiments you should simulate? There is a theoretical answer to that as well, but we are not equipped yet to understand it.[5] But what you can do is you can plot your estimate of the probability $p(H|I)$ as a function of the number of simulated experiments N. If we do that, we will have a visual indication that our estimate is converging. Let's do this.

```
np.random.seed(12345)
N = 10_000 # Yes, no typo here. You can do this with numbers in
↪  Python!
# Simulate the experiments
coin_toss_exp = np.random.randint(0, 2, size=N)
# Approximate the probability
p_H_g_I = coin_toss_exp.sum() / N
# Print the estimate
print(f'p(H|I) ~= {p_H_g_I:1.5f}')
```

We have the experiment done ten thousand times. Now let's produce our estimate of the probability using n experiments for $n = 1$ to $N = 10000$. We can do it with a for loop:

```
# Make an empty array to store the results
p_H_g_I_estimates = np.ndarray((N,))
# Loop over the number of experiments
for n in range(N):
    p_H_g_I_estimates[n] = coin_toss_exp[:n].sum() / (n + 1)
```

You can plot the results with:

```
fig, ax = plt.subplots()
ax.plot(np.arange(1, N + 1), [0.5] * np.ones((N,)), 'r--',
↪  label='True value')
ax.plot(np.arange(1, N + 1), p_H_g_I_estimates, label='Monte Carlo
↪  estimate')
```

---

[5]It's called the *central limit theorem*.

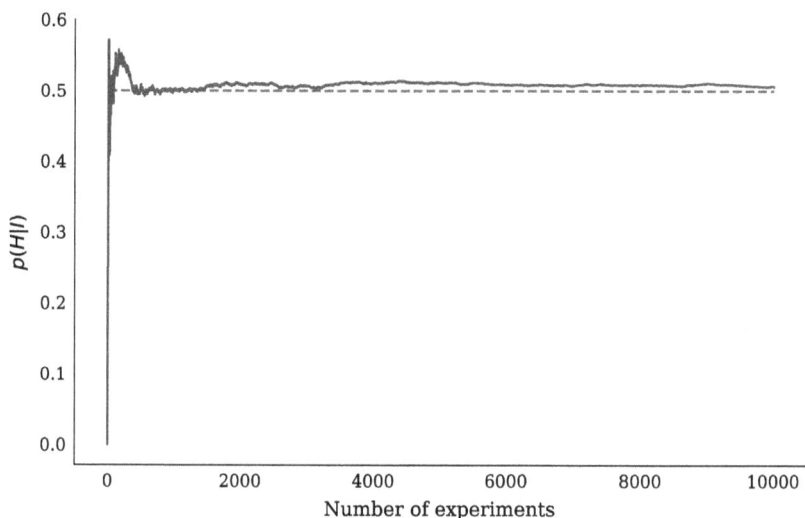

Figure 7.1. The estimate of the probability of heads as a function of the number of experiments simulated.

```
ax.set_xlabel('Number of experiments')
ax.set_ylabel('$p(H|I)$')
save_for_book(fig, 'ch7.fig1')
```

The result is shown in Figure 7.1.

7.5   Estimating Probabilities from Data-Bootstrapping: Colab

You can use the same idea we used in simulations to estimate probabilities from experiments. So, if I is the background information and A is a logical proposition that is experimentally testable, then

$$p(A|I) \approx \frac{\text{Number of times } A \text{ is True under } I \text{ in } N \text{ experiments}}{N}.$$

There is a catch here. The experiments must be *independently* done. This means that you should prepare any apparatus you are using in exactly the same way for all experiments and that no experiment should affect any other in any way. Most of the experiments we run in a lab are independent. However, this assumption may be wrong for data collected in the wild.

7.5.1 Example: Estimating the probability of excessive energy use

Let's try this in practice using the high-performance building dataset. I'm importing the libraries and loading the data below.

```
import numpy as np
import pandas as pd
df = pd.read_excel('temperature_raw.xlsx')
df = df.dropna(axis=0)
df.date = pd.to_datetime(df['date'], format='%Y-%m-%d')
```

The background information I is as follows:

> A random household is picked on a random week during the heating season. The heating season is defined to be the time of the year during which the weekly average of the external temperature is less than 55°F.

The logical proposition A is:

> The weekly HVAC energy consumption of the household exceeds 400 kWh.

First, we start by selecting the subset of the data that pertains to the heating season.

```
df_heating = df[df['t_out'] < 55]
```

We have 2741 such measurements. Now we want to pick a random household on a random week. Unfortunately, in this dataset, this is not exactly equivalent to picking a row at random because there are some missing data. So, if we wanted to be very picky, we would have to find the number of weeks during which we have data from all households. However, we won't be so picky. The result would not be far off what we estimate by randomly picking rows.

Okay, so here is what we are going to do. We will pick N rows at random. For each one of the rows, we are going to test if the logical proposition A is True. Finally, we are going to divide the number of times A is true with N. Alright, let's do it.

```
N = 500
rows = np.random.randint(0, df_heating.shape[0], size=N)
```

We need to pick the rows of `df_heating` that have those indices. We can do it like this:

```
df_heating_exp = df_heating.loc[df_heating.index[rows]]
```

Let's evaluate the value of the logical proposition A for each one of these rows.

```
df_heating_exp_A = df_heating_exp['hvac'] > 300
```

We need to count the number of times the logical proposition was true. We can do this either with the function `pandas.DataFrame.value_counts()`:

```
df_heating_exp_A_counts = df_heating_exp_A.value_counts()
```

If you print it, you will see:

```
hvac
False    463
True      37
Name: count, dtype: int64
```

This returned both the `True` and the `False` counts. To get just the `True` counts:

```
number_of_A_true = df_heating_exp_A_counts[True]
```

And now we can estimate the probability by dividing the number of times A was "True" with the number of randomly selected rows N. We get this:

```
p_A_g_I = number_of_A_true / N
print(f'p(A|I) ~= {p_A_g_I:1.2f}')
```

Output:

```
p(A|I) ~= 0.07
```

Nice! This was easy. Now, you may say why didn't you pick all rows? I could have, and if I had a really, really big number of rows I would be getting a very good estimate of the probability. But you never know if you have enough data. It is not like the simulated experiment where we could do as many runs as we liked. Most of the times, we have a finite amount of data and this is not good enough for an accurate estimate. In other words, there is a bit of epistemic uncertainty in our estimate of the probability. There is something we can do to estimate this epistemic uncertainty. Let me show you.

First, put everything we did above in a nice function:

```
def estimate_p_A_g_I(N, df_heating):
    """Estimates the probability of A given I by randomly picking N
↪   rows
    from the data frame df_heating.

    Arguments:
    N              -      The number of rows to pick at random.
    df_heating     -      The data frame containing the heating data.

    Returns: The number of rows with A True divided by N.
    """
    rows = np.random.randint(0, df_heating.shape[0], size=N)
    df_heating_exp = df_heating.iloc[rows]
    df_heating_exp_A = df_heating_exp['hvac'] > 300
    df_heating_exp_A_counts = df_heating_exp_A.value_counts()
    number_of_A_true = df_heating_exp_A_counts[True] if True in
↪   df_heating_exp_A_counts.keys() else 0
    p_A_g_I = number_of_A_true / N
    return p_A_g_I
```

Now we can call this function as many times as we want. Each time we get an estimate of the probability of A given I. It is going to be a different estimate every time because the rows are selected at random. Here it is 10 times:

```
for i in range(10):
    p_A_g_I = estimate_p_A_g_I(500, df_heating)
    print(f'{i+1:d} estimate of p(A|I): {p_A_g_I:1.3f}')
```

Output:

```
1 estimate of p(A|I): 0.060
2 estimate of p(A|I): 0.070
3 estimate of p(A|I): 0.054
4 estimate of p(A|I): 0.072
5 estimate of p(A|I): 0.060
```

```
6 estimate of p(A|I): 0.064
7 estimate of p(A|I): 0.054
8 estimate of p(A|I): 0.066
9 estimate of p(A|I): 0.078
10 estimate of p(A|I): 0.062
```

Alright, every time we get a different number. To get a sense of the epistemic uncertainty, we can do it many, many times, say 1000 times, and plot a histogram of our estimates. Like this:

```
# A place to store the estimates
p_A_g_Is = []
# The number of rows we sample every time
N = 500
# Put 1000 estimates in there
for i in range(1000):
    p_A_g_I = estimate_p_A_g_I(N, df_heating)
    p_A_g_Is.append(p_A_g_I)
# And now do the histogram
fig, ax = make_full_width_fig()
ax.hist(p_A_g_Is)
```

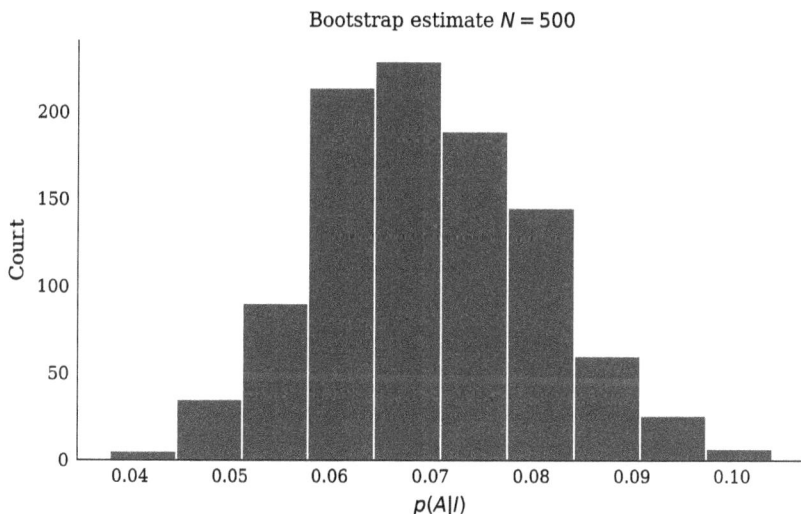

Figure 7.2. The histogram of the estimates of $p(A|I)$ from 1000 bootstrap experiments.

```
ax.set_xlabel('$p(A|I)$')
ax.set_ylabel('Count')
ax.set_title(f'Bootstrap estimate $N = {N:d}$');
save_for_book(fig, 'ch7.fig2')
```

The result is shown in Figure 7.2.

Exercises

1. Blackjack is a popular card game. The background information
   I captures the basic rules of the game relevant to this problem:

   > We have a deck of 52 cards. The deck includes: Four versions
   > aces (A); Four versions of each number from 2 to 10; Four
   > versions of the figures J, Q, and K. In blackjack, all the
   > cards are associated with a number. The cards that have
   > a number on them are associated with that number. The
   > figures J, Q, and K are associated with the number 10.
   > The aces A can either be the number 1 or the number 11.
   > The deck of cards is shuffled adequately.

   Now consider the logical proposition A (blackjack):

   > You draw two cards at random from the deck without
   > replacement. You either have two aces (AA) or the maxi-
   > mum sum of the numbers associated with the cards is 21.
   > For example: (10, A), (J, A), etc.

   Answer the following questions:

   (a) Find the number of ways in which you can choose two unique
       cards from the deck. *Hint*: Google "N choose k."
   (b) Find the number ways in which you can get two cards that
       sum to 21. *Hint*: Enumerate all possibilities.
   (c) Find the probability that you pick two cards that sum to 21,
       i.e., find $p(A|I)$. *Hint*: Use the principle of insufficient reason.

2. We continue with the same background information I and the
   same logical proposition A as in the previous exercise. In this
   problem, you are going to use Monte Carlo simulations to estimate the probability of picking two cards that sum to 21, i.e.,
   $p(A|I)$. Basically, we are going to simulate the process of picking
   these two cards. First, let's start by making all the different cards
   that appear in a deck of 52. In what follows, I use the following
   conventions:

- `'d'` stands for "diamonds."
- `'h'` stands for "hearts."
- `'s'` stands for "spades."
- `'c'` stands for "clubs."

Numbers stand for themselves. And finally:

- `'A'` for "ace."
- `'J'` for "jack."
- `'Q'` for "queen."
- `'K'` for "king."

For example, this is if you see the string `'2h'`, then this is the "two of hearts." If you see `'Ad'`, this is the "Ace of diamonds." And so on. Let's make a deck of cards:

```
deck = []
for n in ['A', '2', '3', '4', '5', '6', '7', '8', '9', '10', 'J',
↪    'Q', 'K']:
    for c in ['d', 'h', 's', 'c']:
        card = n + c
        deck.append(card)
print(deck)
```

We can use `numpy.random.shuffle` to shuffle the deck in place. See the NumPy documentation for more details. Now, let's shuffle the deck:

```
np.random.shuffle(deck)
```

Once the deck is shuffled, you can pick two cards at random by just picking the first two cards of the deck:

```
my_cards = deck[:2]
```

Now, let's write a function that calculates the sum of the cards. I wrote the function so that it only works with two cards. It will always use 11 for aces.

```
def count_cards(cards):
    """"Counts cards according to blackjack conventions.

    Arguments:
    cards    -   Two cards. They much be a string from a deck.

    Returns: The blackjack value of the cards.
```

```
"""
    assert len(cards) == 2, 'This only works for two cards.'
    s = 0
    for c in cards:
        n = c[0]
        if n == 'A':
            s += 11
        elif n == 'J' or n == 'Q' or n == 'K':
            s += 10
        elif len(c) == 3: # this is the case of '10d', '10h',
        ↪   etc.
            s += 10
        else:
            s += int(n)
    return s
```

Let's test it a few times:

```
for i in range(10):
    np.random.shuffle(deck)
    my_cards = deck[:2]
    sum_of_cards = count_cards(my_cards)
    print(my_cards, ' sum to: ', sum_of_cards)
```

(a) Now, we have everything we need. Complete the following code, which uses the Monte Carlo method to estimate the probability of randomly picking two cards that sum to 21. Feel free to experiment with the number of simulations so that you get an accurate estimate.

```
# The number of experiments you want to simulate
num_exp = 1000
# This is a list in which we are going to put the result
# of each "experiment". We will record a 1 (one) if the
↪   experiment
# is successful (cards sum to 21) and a 0 (zero) otherwise
result = []
# Loop over experiments
for i in range(num_exp):
    # YOUR CODE HERE (shuffle the deck)
```

```
my_cards = # YOUR CODE HERE (pick the first two cards
↳   from the deck)
sum_of_cards = # YOUR CODE HERE (find the sum of the
↳   cards)
# YOUR CODE HERE (write a conditional statement that
↳   appends 1
#                   to result if the sum of cards is 21 and
↳   0 otherwise)
p_A_g_I = # YOUR CODE HERE (use result to estimate the
↳ probability of getting two
#                   cards that sum to 21)
print(f'p(A|I) ~= {p_A_g_I:1.5f}')
```

(b) Plot the estimate of $p(A|I)$ as a function of the number of experiments. In the same plot, use a red dashed line to mark the true value of $p(A|I)$ based on your answer to the very first question.

3. **Predicting the probability of major earthquakes in Southern California.** In this exercise, you will predict the probability of major earthquakes in Southern California. We will use the Southern California Earthquake Data Center catalog.[6] The catalog contains all earthquakes recorded from 1932 until now in Southern California. First, let's download the data and put them in a dataframe. On a Jupyter notebook, you can do this by running the following cell:

```
base_url = 'https://raw.githubusercontent.com'
catalog_url = 'SCEDC/SCEDC-catalogs/master/SCEC_DC/'
for year in range(1932, 2021):
    print('Downloading year', year)
    url = f'{base_url}{catalog_url}{year}.catalog'
    !curl -O $url
```

This will download 89 different files, one for each year. It will take about a minute to run. Each one of these is a csv file. We will put them all in the same dataframe for your convenience:

---

[6]https://scedc.caltech.edu/data/cloud.html#eq-catalog.

```
import pandas as pd
list_of_dfs = []
for year in range(1932, 2021):
    filename = '{0:d}.catalog'.format(year)
    print('Reading: ', filename)
    df_year = pd.read_csv(filename, sep=r'\s+', comment='#',
                    names=['Date', 'Hour', 'ET', 'GT', 'MAG',
                    ↪    'M', 'LAT', 'LON',
                            'DEPTH', 'Q', 'EVID', 'NPH',
                            ↪    'NGRM'])
    df_year.Date = pd.to_datetime(df_year['Date'],
    ↪    format='%Y/%m/%d')
    list_of_dfs.append(df_year)
df = pd.concat(list_of_dfs, ignore_index=True)
df['Year'] = pd.DatetimeIndex(df['Date']).year
```

Each row in this dataframe corresponds to an earthquake event
that happened between 1/1/1932 and 12/31/2020. The meaning
of the columns is explained in the SCEDC catalog documentation.
We will only need information from the following columns:

- `Year`: This is the year of the event.
- `ET`: This is the type of the event. There are various types of
  events. For example, the seismometers may pick more than
  earthquakes, e.g., explosions. We are only interested in earthquake events, which are labeled by `eq`.
- `MAG`: This is the magnitude of the event.'

Let's play with the data set to gain some experience. First, let's
extract all data for a random year. Say, year 2019.

```
df_2019 = df[df['Year'] == 2019]
```

Out of these, we only care about earthquake events. So, let's filter
out everything else:

```
df_2019_eq = df_2019[df_2019['ET'] == 'eq']
```

Now, let's see if there was at least one major earthquake during
2019:

```
test_mag = df_2019_eq['MAG'] >= 6
```

Is there at least one True value in this array?

```
print(test_mag.value_counts())
```

There were exactly 2 major earthquakes. You can extract the number like this:

```
number_of_major_earthquakes = test_mag.value_counts()[True]
```

So, to test whether or not there was a major earthquake you need to do:

```
True in test_mag.value_counts().keys()
```

(a) Now, we will use bootstrapping to estimate the probability of a major earthquake during a randomly picked year. Follow the instructions below completing the code where necessary.

```
"""Estimate the probability of major earthquake in a random
↪    year.

Arguments:
num_years     -   The number of years to pick at random.
df            -   The dataframe containing all the observed
↪    events.

Returns: The number of years in which we had at least one
↪    major earthquake divided by the num_years.
"""
num_major_eqs = 0
for i in range(num_years):
    # Pick a year at random between 1932 and 2020
    y = np.random.randint(1932, 2021)
    # Extract all the events that happened in that year
    df_y = # YOUR CODE HERE
    # Find all earthquake events
    df_y_eq = # YOUR CODE HERE
    # Test if there is at least one major earthquake in this
    ↪    year
    test_mag = # YOUR CODE HERE
    test_mag_counts = test_mag.value_counts()
    # Test if there is at least one major event in this year
    # and increase num_major_eqs by one if yes
    if True in test_mag.value_counts():
        num_major_eqs += 1
return num_major_eqs / num_years
```

Use the following lines to test your code. We run it 10 times. Notice that every time you get a slightly different estimate.

```
for i in range(10):
    p_major_eq =
    ↪ estimate_probability_of_major_earthquake_during_year(50,
    ↪ df)
    print(f'p_major_eq = {p_major_eq:1.2f}')
```

(b) Repeat the probability estimation above 200 times, store all estimates in a list, and do a histogram of the estimates.

```
# A place to store the estimates
p_major_eqs = []
# Put 1000 estimates in there
for i in range(200):
    print(i)
    p_major_eq = # your code here
    p_major_eqs.append(p_major_eq)
# And now do the histogram
# Your code here
```

Chapter 8

The Basic Rules of Probability

In this chapter, I introduce you to the fundamental rules of probability theory that form the foundation of statistical reasoning and data analysis. I start with the "obvious rule" of probability, which establishes the basic principles of probability assignments. Then, I explore the product rule, also known as Bayes' rule, which is crucial for understanding conditional probabilities and updating beliefs in light of new evidence. Finally, I examine the sum rule, which helps us calculate probabilities by breaking them down into simpler components.

8.1 The Basic Rules of Probability

There are two rules of probability from which everything else can be derived from the desiderata we introduced in Chapter 7. They are as follows:

- The **obvious rule**:

$$p(A|I) + p(\neg A|I) = 1.$$

- The **product rule** (or Bayes' rule):

$$p(AB|I) = p(A|BI)p(B|I).$$

   I call it the first rule, the "obvious rule" because it states the obvious: A logical proposition about the world is either True or False.

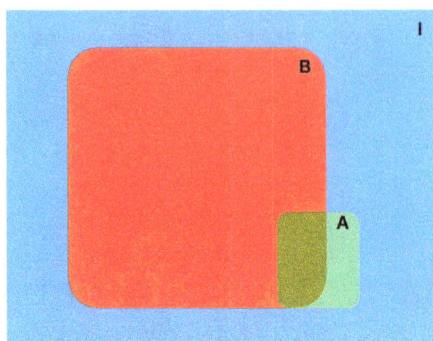

Figure 8.1. Venn diagram showing the information I, and the logical propositions A and B.

Let me stress again that it is vitally important that you do not try to apply probability in a system that includes contradictions.

The product rule is not obvious. It states that the probability of the logical propositions A and B being True at the same time is the probability of A being True given that B is True times the probability that B is True. Understanding it requires a bit of meditation. We can use the Venn diagram of Figure 8.1 to develop some intuition about it.

You can think of a Venn diagram as a 2D representation of everything that can happen in the world. Each point inside the blue area is a possibility allowed by the information I. The red area marks the possibilities in which B is True and the green area marks the possibilities in which A is True. The intersection of the red and the green area (brown) marks the possibilities in which both A and B are True.

In Venn diagrams, probabilities correspond to normalized areas. It goes like this:

$$p(\text{smth}|\text{smth else}) = \frac{\text{area of intersection of smth and smth else}}{\text{area of smth else}}.$$

Strictly speaking, this is not an equality but a correspondence or analogy. In our particular case, we have,

$$p(B|I) = \frac{\text{area of intersection of } B \text{ and } I \text{ (red)}}{\text{area of } I \text{ (blue)}}$$

$$= \frac{\text{area of } B \text{ (red)}}{\text{area of } I \text{ (blue)}}.$$

Similarly,

$$p(A|BI) = \frac{\text{area of the intersection of } A \text{ and B and } I \text{ (brown)}}{\text{area of intersection of } B \text{ and } I}$$

$$= \frac{\text{area of the intersection of } A \text{ and B (brown)}}{\text{area of } B \text{ (red)}},$$

and

$$p(AB|I) = \frac{\text{area of intersection of } A \text{ and B (brown)}}{\text{area of } I \text{ (blue)}}.$$

Thus, we have,

$$p(AB|I) = \frac{\text{area of intersection of } A \text{ and B (brown)}}{\text{area of } I \text{ (blue)}}$$

$$= \frac{\text{area of intersection of } A \text{ and B (brown)}}{\text{area of } I \text{ (blue)}} \cdot \frac{\text{area of } B \text{ (red)}}{\text{area of } B \text{ (red)}}$$

$$= \frac{\text{area of intersection of } A \text{ and B (brown)}}{\text{area of } B \text{ (red)}} \cdot \frac{\text{area of } B \text{ (red)}}{\text{area of } I \text{ (blue)}}$$

$$= p(A|BI)p(B|I),$$

which is the product rule. Keep in mind that this is not a mathematical proof. However, it can help you develop a little bit of intuition about the product rule.

8.1.1 *Example: Drawing balls from a box without replacement*

Consider the following information I:

We are given a box with 10 balls, 6 of which are red and 4 of which are blue (Figure 8.2). The box is sufficiently mixed so that when we get a ball from it, we don't know which one we pick. When we take a ball out of the box, we do not put it back.

Now, let's say that we draw the first ball. Let B_1 be the logical proposition:

The first ball we draw is blue.

Figure 8.2.   A box with 10 balls.

What is the probability of B_1? Our intuition tells us to set:

$$p(B_1|I) = \frac{4}{10} = \frac{2}{5}.$$

This is known as the *principle of insufficient reason.*

We can now use the **obvious rule** to find the probability of drawing a red ball, i.e., of $\neg B_1$. Of course, $\neg B_1$ is just the sentence:

> The first ball we draw is red.

So, let's also call $\neg B_1$ with the name R_1. It is

$$p(R_1|I) = p(\neg B_1|I) = 1 - p(B_1|I) = 1 - \frac{2}{5} = \frac{3}{5}.$$

Great! Let's continue with drawing a second ball. Consider the sentence R_2:

> The second ball we draw is red.

What is the probability of R_2 given that B_1 is true? We just need to use common sense to find this probability:

- We had 10 balls, 6 red and 4 blue.
- Since B_1 is true (the first ball was blue), we now have 6 red and 3 blue balls.
- Therefore, the probability that we draw a red ball next is:

$$p(R_2|B_1, I) = \frac{6}{9} = \frac{2}{3}.$$

Similarly, we can find the probability that we draw a red ball in the second draw, given that we drew a red ball in the first draw:

- We had 10 balls, 6 red and 4 blue.
- Since R_1 is true (the first ball is red), we now have 5 red and 4 blue balls.
- Therefore, the probability that we draw a red ball next is:

$$p(R_2|R_1, I) = \frac{5}{9}.$$

All, this was easy. Let's find the probability that we draw a blue ball in the first draw and a red ball in the second draw. This time, we have to use the **product rule**:

$$p(B_1, R_2|I) = p(R_2|B_1, I)p(B_1|I) = \frac{2}{3}\frac{2}{5} = \frac{4}{15}.$$

8.2 Probability of Logical Disjunctions

All other rules of probability theory can be derived from the two basic rules. So take two logical propositions A and B. The logical disjunction of A and B is the logical proposition $(A \text{ or } B)$. In our notations, we write $A + B$ for $(A \text{ or } B)$. We can deduce from the basic rules of probability that:

$$p(A + B|I) = p(A|I) + p(B|I) - p(AB|I).$$

In words, this equation says that the probability of A or B being True is the probability that A is True plus that probability that B is True minus the probability that both A and B are True. This statement is very easy to understand intuitively by looking again at the Venn diagram in Figure 8.1. The probability $p(A + B|I)$ is the area of the union of A with B (normalized by I). This area is indeed the area of A (normalized by I) plus the area of B (normalized by I) minus the area of A and B (normalized by I), which was double-counted.

Alternatively, we can provide a formal proof of this equation that relies only on the two basic rules of probability. Here it is

$$p(A + B|I) = 1 - p(\neg(A + B)|I)$$
$$= 1 - p(\neg A, \neg B|I) \text{ (obvious rule)}$$

$$= 1 - p(\neg A|\neg B, I)p(\neg B|I) \text{ (product rule)}$$
$$= 1 - [1 - p(A|\neg B, I)]\, p(\neg B|I) \text{ (obvious rule)}$$
$$= 1 - p(\neg B|I) + p(A|\neg B, I)p(\neg B|I)$$
$$= 1 - p(\neg B|I) + p(A\neg B|I) \text{ (product rule)}$$
$$= 1 - p(\neg B|I) + p(\neg B|A, I)p(A|I) \text{ (product rule)}$$
$$= 1 - p(\neg B|I) + [1 - p(B|A, I)]\, p(A|I) \text{ (obvious rule)}$$
$$= 1 - p(\neg B|I) + p(A|I) - p(B|A, I)p(A|I)$$
$$= 1 - [1 - p(B|I)] + p(A|I) - p(B|A, I)p(A|I)$$

(obvious rule)

$$= p(A|I) + p(B|I) - p(B|A, I)p(A|I)$$
$$= p(A|I) + p(B|I) - p(AB|I) \text{ (product rule)}.$$

8.3   The Sum Rule

This is the final rule we are going to consider in this lecture. It is one of the most important rules. **You absolutely have to memorize it.** It goes as follows.

Consider the sequence of logical sentences $B_1, \ldots, B_n$ such that:

- One of them is definitely true:

$$p(B_1 + \cdots + B_n|I) = 1.$$

- They are mutually exclusive:

$$p(B_i B_j|I) = \delta_{ij} = \begin{cases} 1, & \text{if } i = j, \\ 0, & \text{otherwise.} \end{cases}$$

Then, for any logical sentence A, we have

$$p(A|I) = \sum_{i=1}^{n} p(AB_i|I) = \sum_{i=1}^{n} p(A|B_i, I)p(B_i|I).$$

Again, this requires a bit of meditation. You take any logical sentence A and set of mutually exclusive and exhaustive possibilities

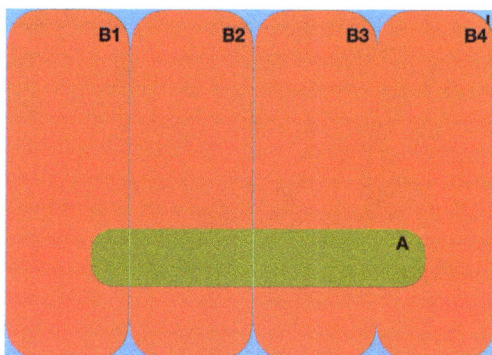

Figure 8.3. Visualization of the sum rule of probability.

$B_1, \ldots, B_n$ and you break down the probability of A in terms of the probabilities of the B_i's. The Venn diagrams in Figure 8.3 can help you understand what is going on.

The sum rule can be trivially proved by induction using only the obvious rule and the product rule. It is instructive to go through the proof. For $n = 2$, we have

$$p(A|I) = p(A \text{ and } (B_1 \text{ or } B_2)|I)$$
$$= p\left((A \text{ and } B_1) \text{ or } (A \text{ and } B_2)|I\right)$$
$$= p(A \text{ and } B_1|I) + p(A \text{ and } B_2|I) - p\left((A \text{ and } B_1) \text{ and } (A \text{ and } B_2)|I\right)$$
$$= p(AB_1|I) + p(AB_2|I) - p(AB_1B_2|I)$$
$$= p(AB_1|I) + p(AB_2|I),$$

because

$$p(AB_1B_2|I) = p(B_1B_2|I)p(A|I) \leq p(B_1B_2|I) = 0.$$

And then, assume that it holds for n, you can easily show that it also holds for $n + 1$ completing the proof.

8.3.1 Example: Drawing balls from a box without replacement (sum rule)

We now continue our analysis of Section 8.1.1. Let us consider the probability of getting a red ball in the second draw without observing in the first draw $p(B_1|I)$. We have two possibilities for the first draw. We either got a blue ball (B_1 is true) or we got a red ball (R_1 is true). In other words, B_1 and R_1 cover all possibilities and are mutually exclusive. We can use the sum rule:

$$p(R_2|I) = p(R_2|B_1, I)p(B_1|I) + p(R_2|R_1, I)p(R_1|I)$$

$$= \frac{2}{3}\frac{2}{5} + \frac{5}{9}\frac{3}{5}$$

$$= 0.6.$$

8.3.2 *Example: Drawing balls from a box without replacement (information flow)*

There is one more point that I would like to make about the urn example. If you paid close attention, in all our examples the conditioning we did followed the causality. For instance, in the urn example we where writing $p(R_2|B_1, I)$ for the probability of getting a red ball in the second draw after having observed the blue ball in the first draw. However, conditioning on stuff **does not have to follow the causal links**. It is completely legitimate to ask what is the probability of a blue ball in the first draw, given that you have observed that the result of the second draw is a red ball.

That is, you can write down the mathematical expression $p(B_1|R_2, I)$. This does not mean that R_2 is causing B_1. What happens here is that observing R_2 changes your state of knowledge about B_1. This is an example of information flowing in the reverse order of a causal link. Let's solve it analytically. We have by applying the product rule:

$$p(B_1|R_2, I) = \frac{p(B_1, R_2|I)}{p(R_2|I)}$$

$$= \frac{\frac{4}{15}}{0.6}$$

$$\approx 0.44.$$

This is greater than the probability of drawing a blue ball in the first place, $p(B_1|I) = 0.4$. Does this make sense? Yes it does! Here is how you should think:

- You draw a ball without seeing it and you put it in another box.
- You draw the second ball and you see that it is a red one.
- This means that this particular red ball was not picked in the first draw.
- So, it is as if in the first draw you had one less red to worry about which increases the probability of a blue.
- So, it is as if you had 5 red balls and 4 blue balls giving you a probability of blue $\frac{4}{9} \approx 0.44$.

This is amazing! It agrees perfectly with the prediction of the product rule. This was one of our desiderata (if you compute something in two different ways you should get the same result). You can rest assured that as soon as you use the product rule and the sum rule and logic, it is impossible to get the wrong answer. That is, if you can actually carry out the computations.

Exercises

(1) **Space Station Fault Detection Analysis**[1] We are tasked with assessing the reliability of a fault detection algorithm that has been deployed on the International Space Station. The algorithm is designed to detect a refrigerant leak in the cooling system of the station. The prior information I is

> The percentage of refrigerant systems that develop leaks during a mission is 0.4%. We have run several experiments and determined that:
>
> - If a system has a leak, then 80% of the time the detection system correctly identifies it.
> - If a system is functioning normally, then 90% of the time the detection system correctly reports no leak.

To facilitate your analysis, consider the following logical sentences concerning a system:

> A: The system is monitored and the detection system reports a leak. B: The system has an actual leak.

---

[1]This example is a modified version of the one found in a 2013 lecture on Bayesian Scientific Computing taught by Prof. Nicholas Zabaras.

(a) Find the probability that the system has a fault (before looking at the detection system's report), i.e., $p(B|I)$. This is known as the base rate or the prior probability.

(b) Find the probability that the detection system reports a leak given that the system has an actual leak, i.e., $p(A|B,I)$.

(c) Find the probability that the detection system does not report a leak given that the system does not have an actual leak, i.e., $p(\neg A|\neg B, I)$.

(d) Find the probability that the system has an actual leak given that the detection system reports a leak, i.e., $p(B|A, I)$.

(e) Find the probability that the system has an actual leak given that the detection system does not report a leak, i.e., $p(B|\neg A, I)$. Does the detection system change our prior state of knowledge about the system? Is the detection system useful?

(f) What would a good detection system look like? Find values for

$$p(A|B, I) = p(\text{detection system reports leak}|\text{leak}, I),$$

and

$$p(A|\neg B, I) = p(\text{detection system reports leak}|\text{no leak}, I),$$

so that

$$p(B|A, I) = p(\text{leak}|\text{detection system reports leak}, I) = 0.99.$$

There are more than one solutions. How would you pick a good one? Thinking in this way can help you set goals if you work in R&D. If you have time, try to figure out whether or not such an accurate detection system is feasible for the space station.

Chapter 9

Discrete Random Variables

In this chapter, I introduce you to discrete random variables and their probability mass functions. We explore how the Bernoulli distribution models experiments with two possible outcomes, and I demonstrate the Categorical distribution using a familiar example of a six-sided die.

9.1 Definition of Random Variables

Imagine that you are conducting an experiment, e.g., a coin toss, throwing a die. How can you talk about the result of the experiment before you actually do it? This is where the concept of a *random variable* comes to the rescue. You can simply use a letter like X to indicate the result of the experiment before you actually do it. So, remember:

> A random variable X models the result of a random experiment.[1]

Now, if the result of the experiment consists of discrete labels, we say that we have a *discrete random variable*. The result of a coin toss is a discrete random variable because there are only two possibilities

---

[1]There is a precise mathematical definition of random variables. However, it requires introducing some advanced concepts (probability spaces, measurable functions, etc.), which are all well beyond the scope of this book. We should be fine with our informal definition of a random variable within this book.

(heads and tails). The result of throwing a six-sided die is also a discrete random variable because there are six possibilities.

However, there are many experiments that do not have discrete outcomes. For example, consider the experiment that measures the mass of a manufactured steel ball. This is a scalar number. Such experiments require *continuous random variables*, which will be the subject of Chapter 10. From now on, we will restrict our attention to discrete random variables.

Discrete random variables can really take any type of value. For example, the coin toss results in "heads" or "tails." The six-sided die in 1, 2, 3, 4, 5, or 6. Other possibilities of discrete random variable values are colors: "red," "yellow," "green," etc. There is no end to this. However, since these types have discrete labels you can always code them using natural numbers $0, 1, 2, 3, \ldots$ For example, when we are dealing with a coin toss, we can code "heads" with 0 and "tails" with 1. Or when we are dealing with color labels, we can code "red" with 0, "yellow" with 1 and "green" with 2. So, there is no loss of generality in assuming that the result of an experiment resulting in a discrete random variable is some number $0, 1, 2, \ldots$. Finally, notice that we allow for an infinite number of possibilities as long as they remain discrete.

9.2   The Probability Mass Function

Say that X is the result of a coin toss experiment. As we have already discussed with can code the result of the experiment with natural numbers. Here, we use 0 for "heads" and 1 for "tails." If we have a fair coin, we know that X takes the value 0 with probability 0.5 and the value 1 also with probability 0.5. We can write this succinctly as

$$X = \begin{cases} 0, & \text{with probability } 0.5, \\ 1, & \text{otherwise.} \end{cases} \tag{9.1}$$

But there is another way to write this that is often more convenient. Here it is

$$p(X - 0) - \text{probability that } X \text{ takes the value } 0 = 0.5. \tag{9.2}$$

We can use the obvious rule of probability to get:

$$p(X = 1) = 1 - p(X \neq 1) = 1 - p(X = 0) = 1 - 0.5 = 0.5. \quad (9.3)$$

You can now define the following function:

$$f_X(x) := \text{probability that } X \text{ takes the value } x \equiv p(X = x). \quad (9.4)$$

This is a function associated with the discrete random variable X (that is why X is used as a subscript in f_X). It takes any possible value x for X and responds with the probability of X taking that value. It has a special name. It is called the *probability mass function* or PMF of the discrete random variable X.

We have silently introduced the following standard notational conventions:

- Use upper case letters to represent random variables, like X, Y, Z.
- Use lower case letters to represent the values of random variables, like x, y, z can take.

The notation $f_X(x)$ for the PMF is a bit of an overkill in my opinion. It is commonly used in mathematical textbooks where rigor is valued above everything. In practice, no one wants to write or type this many symbols. So, it is a common convention to write $p(x)$ instead of $f_X(x)$ when there is no ambiguity. Symbolically, we follow this convention:

$$p(x) \equiv f_X(x) \equiv p(X = x). \quad (9.5)$$

In other words, when we write $p(x)$, we are talking about the PMF of a random variable capitalize$(x) = X$ evaluated at x. Similarly, if we write $p(y)$ we are referring to the PMF of the random variable Y evaluated at y. However, it is meaningless to write $p(0)$. What does it mean? To which random variable do we refer? X or Y? So, in case we want to plug in a specific value, we should write $p(X = 0)$ or $p(Y = 0)$, etc.[2]

---

[2]In everything I write below, there is supposed to be some background information I. So, I should be really writing $p(X = 0|I), p(x|I)$, etc. However, since I remains fixed I decided to drop it from the notation.

9.3   The Bernoulli Distribution: Colab

I am now going to generalize the coin toss experiment. Imagine an experiment with two outcomes: 0 or 1. You can think of 0 as "failure" and 1 as "success." Assume that the experiment is successful with probability θ, which is some number between 0 and 1. For example, for a fair coin $\theta = 0.5$. But we allow for any value. The result of such an experiment is captured by the following random variable:

$$X = \begin{cases} 1, & \text{with probability } \theta, \\ 0, & \text{otherwise.} \end{cases} \tag{9.6}$$

In terms of the probability mass function, we have

$$p(X = 0) = \theta, \tag{9.7}$$

and

$$p(X = 1) = 1 - \theta. \tag{9.8}$$

This random variable has a special name. It is called a Bernoulli random variable, named after Jacob Bernoulli, a famous Swiss mathematician who lived in the 17th century. Another way to say that X is a Bernoulli random variable is to write:

$$X \sim \text{Bernoulli}(\theta). \tag{9.9}$$

which is read as

> The random variable X *follows* a Bernoulli *distribution* with parameter θ.

Let's use the functionality of the Python module `scipy.stats` to define a Bernoulli random variable and sample from it. I will pick $\theta = 0.6$.

```
import scipy.stats as st
theta = 0.6
X = st.bernoulli(theta)
```

The object `X` encapsulates everything related to the random variable X. For example, `X.support()` tells you which values it takes:

```
X.support()
```

The result is a tuple of two numbers:

```
(0, 1)
```

This means that X can take the values 0 and 1, as expected. To evaluate the probability mass function, say at $X = 0$, you can use:

```
X.pmf(0)
```

Finally, you can use `X.rvs()` to generate samples from this random variable.

```
for i in range(10):
    print(X.rvs())
```

This will print 10 random samples from the Bernoulli distribution with parameter $\theta = 0.6$. If you want to put the samples in an array, you can use:

```
samples = X.rvs(1000)
```

Let's now show the histogram of the samples:

```
fig, ax = plt.subplots()
ax.hist(samples, bins=2, density=True)
plt.show()
```

The result is shown in Figure 9.1.

9.4   Properties of the Probability Mass Function

There are some standard properties of the probability mass function that are worth memorizing. First, the probability mass function is nonnegative:

$$p(x) \geq 0,$$

for the possible values x that the random variable X can take. Second, the probability mass function is normalized:

$$\sum_x p(x) = 1.$$

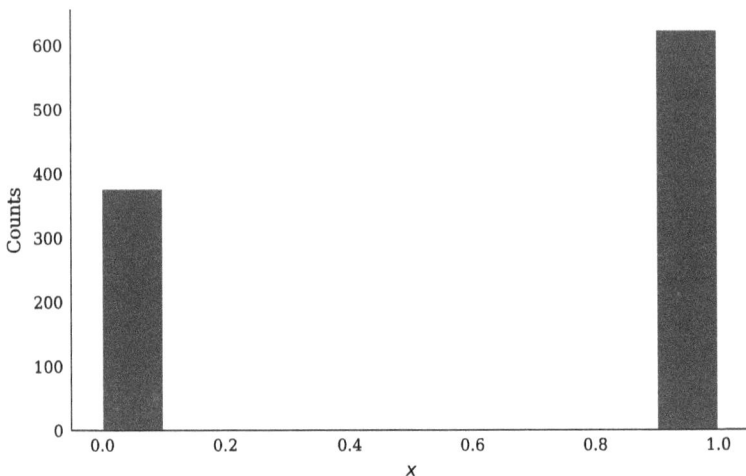

Figure 9.1. Histogram of 1000 samples from the Bernoulli distribution with parameter $\theta = 0.6$.

The sum is over all the possible values of X. This is a direct consequence of the fact that X must take some value of all that are possible.

Finally, you can use the probability mass function to find the probability that X takes values in any set of possibilities. For example,

$$p(X = x_1 \text{ or } X = x_2) = p(x_1) + p(x_2),$$

if $x_1 \neq x_2$. Similarly, if x_1, x_2, and x_3 are all different, then:

$$p(X = x_1 \text{ or } X = x_2 \text{ or } X = x_3) = p(x_1) + p(x_2) + p(x_3).$$

And, in general for any set of possible values of X, say A, the probability of X taking values in A is

$$p(X \in A) = \sum_{x \in A} p(x).$$

9.5   The Categorical Distribution: Colab

We are now going to generalize the six-sided die experiment. A Categorical random variable is used to model an experiment with taking

K different possibilities coded, for example, $1, 2, \ldots, K$, each with a different probability. We can write:

$$X = \begin{cases} 1, & \text{with probability } p_1, \\ 2, & \text{with probability } p_2, \\ \vdots \\ K, & \text{with probability } p_K. \end{cases}$$

Of course, we can also write:

$$p(X = x) = p_x.$$

Another way, we can write this is

$$X \sim \text{Categorical}(p_1, \ldots, p_K),$$

which we read as

> The random variable X follows a Categorical distribution with K possibilities each with probability p_1, p_2 to p_K.

The six-sided, fair, die is a particular example of a Categorical. This one in particular:

$$p(X = x) = \frac{1}{6},$$

if $x = 1, 2, \ldots, 6$.

Let's now make a specific choice for the probabilities, make a Categorical, and sample from it. We are going to play with this one which has four possibilities:

$$X \sim \text{Categorical}(0.1, 0.3, 0.4, 0.2).$$

Here is the code:

```
# The probabilities:
ps = [0.1, 0.3, 0.4, 0.2] # this has to sum to 1
# And here are the corresponding values:
xs = np.array([1, 2, 3, 4])
# Here is how you can define a categorical rv:
X = st.rv_discrete(name='Custom Categorical', values=(xs, ps))
```

You can evaluate the PMF at any value of X using:

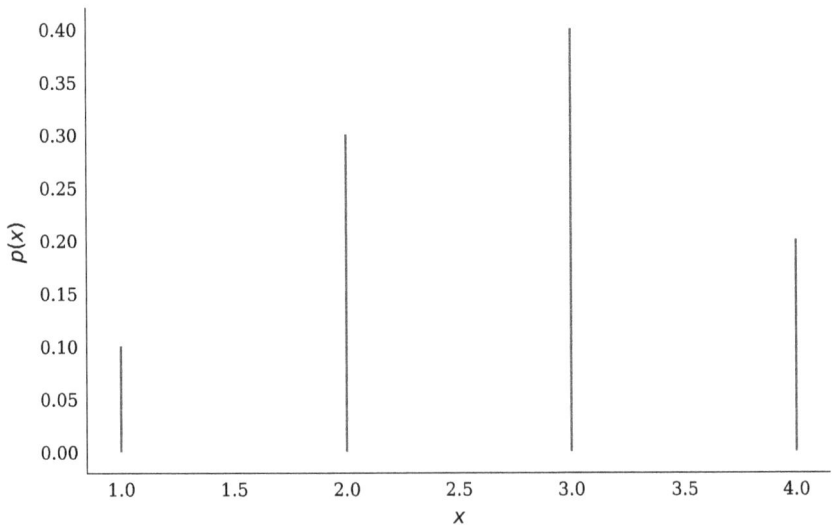

Figure 9.2. PMF of the Categorical distribution with parameters $p_1 = 0.1, p_2 = 0.3, p_3 = 0.4, p_4 = 0.2$.

```
X.pmf(2) # gives 0.3
```

And you can sample from it like this:

```
X.rvs(size=10)
```

A possible result looks like this:

```
array([3, 3, 3, 4, 3, 4, 4, 4, 3, 2])
```

Let's plot the PMF:

```
fig, ax = plt.subplots()
ax.vlines(xs, 0, X.pmf(xs))
plt.show()
```

The result is shown in Figure 9.2.

Now let's find the probability that X takes the value 2 or 4. It is

$$p(X = 2 \text{ or } X = 4) = p(X = 2) + p(X = 4).$$

With Python code:

```
X.pmf(2) + X.pmf(4)
```

The result is

```
0.5
```

9.6 The Binomial Distribution: Colab

Suppose that we are dealing with an experiment with two outcomes 0 (failure) and 1 (success) and that the probability of success is θ. We are interested in the random variable X that counts the number of successful experiments in n trials. This variable is called a Binomial random variable. We write:

$$X \sim \text{Binomial}(n, \theta).$$

It can be shown using combinatorics, that the probability of k successful experiments is given by the PMF:

$$p(X = k) = \binom{n}{k} \theta^k (1 - \theta)^{n-k},$$

where $\binom{n}{k}$ is the number of k combinations out of n elements, i.e.,

$$\binom{n}{k} = \frac{n!}{k!(n-k)!}.$$

Let's now define the binomial in `scipy.stats`:

```
n = 5        # Performing the experiment n times
theta = 0.6 # Probability of success each time
X = st.binom(n, theta) # Number of successes
```

Here are some samples:

```
X.rvs(10)
```

A possible result is

```
array([2, 4, 3, 4, 3, 4, 3, 2, 3, 3])
```

The PMF is shown in Figure 9.3.

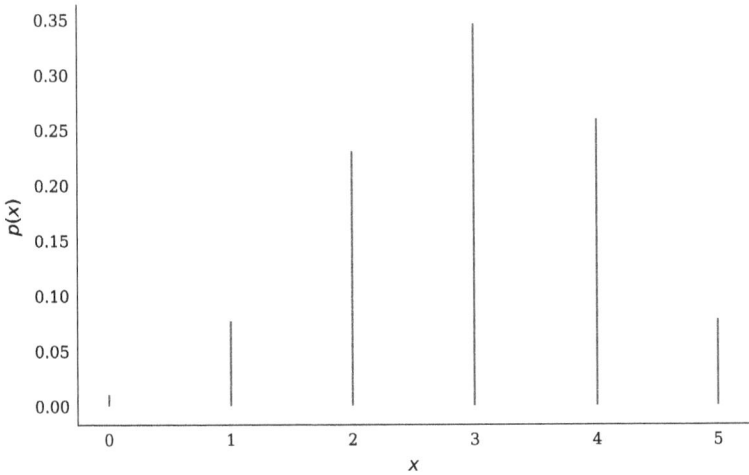

Figure 9.3. PMF of the Binomial distribution with parameters $n = 5$ and $\theta = 0.6$.

9.7 The Poisson Distribution: Colab

The Poisson distribution models the number of times an event occurs in an interval of space or time. For example, a Poisson random variable X may be:

- The number earthquakes greater than 6 Richter occurring over the next 100 years.
- The number of major floods over the next 100 years.
- The number of patients arriving at the emergency room during the night shift.
- The number of electrons hitting a detector in a specific time interval.

The Poisson is a good model when the following assumptions are true:

- The number of times an event occurs in an interval takes values $0, 1, 2, \ldots$.
- Events occur independently.
- The probability that an event occurs is constant per unit of time.

- The average rate at which events occur is constant.
- Events cannot occur at the same time.

When these assumptions are valid, we can write:

$$X \sim \text{Poisson}(\lambda), \tag{9.10}$$

where $\lambda > 0$ is the rate with each the events occur. You read this as

> The random variable X follows a Poisson distribution with rate
> parameter λ.

The PMF of the Poisson is

$$p(X = k) = \frac{\lambda^k e^{-\lambda}}{k!}. \tag{9.11}$$

Let's look at a specific example. Historical data show that at a given region a major earthquake occurs once every 100 years on average. What is the probability that k such earthquakes will occur within the next 100 years. Let X be the random variable corresponding to the number of earthquakes over the next 100 years. Assuming the Poisson model is valid, the rate parameter is $\lambda = 1$ and we have

$$X \sim \text{Poisson}(1). \tag{9.12}$$

Here is a Poisson random variable in `scipy.stats`:

```
X = st.poisson(1.0)
```

Here are some samples:

```
X.rvs(10)
```

A possible result is:

```
array([0, 0, 1, 1, 1, 2, 3, 0, 0, 4])
```

I show the PMF in Figure 9.4. The support of the Poisson distribution is the set of non-negative integers. But I stop the plot at $X = 5$ because the probability of $X > 5$ is very small.

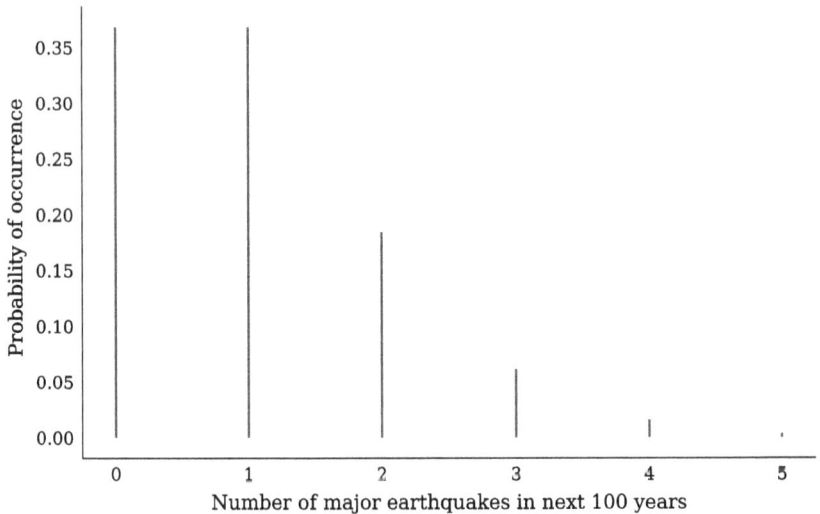

Figure 9.4. PMF of the Poisson distribution with rate parameter $\lambda = 1$.

Exercises

1. Sample from the Bernoulli distribution with parameter $\theta = 0.8$ and plot the histogram.

2. Define the Categorical random variable $X \sim \text{Categorical}(0.1, 0.1, 0.4, 0.2, 0.2)$ taking values $1, 2, 3, 4$ and 5. Plot the PMF of X. Evaluate the probability that X takes the values 2, 3 or 5.

3. Create a binomial random variable with $n = 20$ and $\theta = 0.6$. Plot the PMF of X. Repeat the experiment for $n = 100$. How does the resulting PMF look like? It starts to look like a bell curve!

4. How would the rate parameter λ change if the rate with each major earthquakes occurred in the past was 2 every 100 years? Plot the PMF of the new Poisson random variable. You may have to add more points in the x-axis.

5. **Predicting the probability of major earthquakes in Southern California.** We are going to revisit Exercise 3 from Chapter 7, but this time we are going to use a Poisson distribution to carry out our analysis.

The San Andreas fault[3] extends through California forming the boundary between the Pacific and the North American tectonic plates. It has caused some of the major earthquakes on Earth. We are going to focus on Southern California and we would like to assess the probability of a major earthquake, defined as an earthquake of magnitude 6.5 or greater, during the next 10 years.

The first thing we are going to do is go over a database of past earthquakes[4] that have occurred in Southern California and collect the relevant data. We are going to start at 1900 because data before that time may be unreliable. I have gone over each decade and counted the occurrence of a major earthquake (i.e., count the number of orange and red colors in each decade). Here are the data:

```
eq_data = np.array([
    0, # 1900-1909
    1, # 1910-1919
    2, # 1920-1929
    0, # 1930-1939
    3, # 1940-1949
    2, # 1950-1959
    1, # 1960-1969
    2, # 1970-1979
    1, # 1980-1989
    4, # 1990-1999
    0, # 2000-2009
    2 # 2010-2019
])
```

Let's use the Poisson to model the number of earthquakes X occurring in a decade. We write

$$X \sim \text{Poisson}(r),$$

where r is the *rate parameter* of Poisson. The rate is the number of events per time period. Here, r is the number of earthquakes per decade.

---

[3]https://en.wikipedia.org/wiki/San_Andreas_Fault.
[4]https://scedc.caltech.edu/significant/chron-index.html.

(a) Use the `eq_data` to find the rate r of the Poisson. We can set it as the empirical average of the observed number of earthquakes per decade.

(b) Initialize a Poisson random variable X with rate parameter r using `scipy.stats`.

(c) Plot the PMF of X.

(d) What is the probability that no major earthquake will occur during the next decade?

(e) What is the probability that one or two major earthquakes will occur during the next decade?

(f) What is the probability that at least one major earthquake will occur during the next decade?

Chapter 10

Continuous Random Variables

In this chapter, I introduce you to continuous random variables and their probability density functions (PDFs). I'll show you how probabilities correspond to areas under the PDF curve and demonstrate how histograms converge to the true probability distribution as we collect more samples and use finer bin sizes.

The definition of a continuous random variable is pretty straightforward:

> A *continuous random variable* models the result of an experiment that takes continuous values.

There are a few mathematical concepts that we need to introduce in order to play with continuous random variables. These are probability density functions, cumulative distribution functions, expectations and variances. We will build these concepts gradually.

10.1 Example: Uncertainties in Steel Ball Manufacturing

Steel balls are an essential component of ball bearings, which, in turn, are an essential component of axles in bikes, cars, etc. It is very important to tightly control the properties of the ball, e.g., diameter and steel density. Any slight variation increases undesired vibrations and as a consequence results in noise and faster degradation of ball bearings.

How do you characterize the manufacturing uncertainty on steel ball properties? Well, you can use continuous random variables for that. For example X could be the random variable that gives you the diameter of a ball (assuming they are perfectly spherical — they are not).

Assume that we know that the balls have a diameter that is greater than 0.95 mm and smaller than 1.05 mm. What sort of random variable should we assign to the diameter? Let's call this random variable X. We know a few things right away about it. First, X cannot be smaller than 0.95 mm. So,

$$p(X < 0.95 \text{ mm}) = 0.$$

Second, X cannot be greater than 1.05 mm. So,

$$p(X > 1.05 \text{ mm}) = 0.$$

Then, we also know that X should be between 0.95 mm and 1.05 mm. That is:

$$p(0.95 \text{ mm} \leq X \leq 1.05 \text{ mm}) = 1.$$

All this is trivial. Is there anything non-trivial we can say about X? Take a diameter x from 0.95 mm to 1.05 mm and consider a small window Δx around it. What is the probability that X takes values between x and $x + \Delta x$? Mathematically, what is the probability $p(x \leq X \leq x + \Delta x)$? For this particular example, it makes sense to assume that $p(x \leq X \leq x + \Delta x)$ is proportional to Δx:

$$p(x \leq X \leq x + \Delta x) = c\Delta x,$$

where $c > 0$ is a proportionality constant. The bigger Δx is, the higher the probability. What is the proportionality constant? You can find the proportionality constant by considering an extreme case. Pick $x = 0.95$ mm and $\Delta x = 0.1$ mm. Then:

$$c \cdot 0.1 \text{ mm} = p(0.95 \text{ mm} \leq X \leq 1.05 \text{ mm}) = 1,$$

from which we get that:

$$c = 10 \text{ mm}^{-1}.$$

The constant c is some sort of density. It gives you how much prob ability you get per unit of x. In this particular case, the probability density is just a constant (i.e., it does not depend on x). However, in general it may be depend on x and then we call it a *probability density function* or PDF. We will talk about probability density functions in the next section.

If you have the probability density, you can find any probability regarding X you want. For example, the probability that X is between 1 mm and 1.03 mm is

$$p(1 \text{ mm} \leq X \leq 1.03 \text{ mm}) = p(1 \text{ mm} \leq X \leq 1 \text{ mm} + 0.03 \text{ mm})$$

$$= c\Delta x$$

$$= 10 \text{ mm}^{-1} \cdot 0.03 \text{ mm}$$

$$= 0.3.$$

Here is another one. Say that you want to find the probability that X is smaller than 0.99 mm or greater than 1.03 mm. How do you do that? Well, here it is

$$p(X < 0.99 \text{ mm or } X > 1.03 \text{ mm})$$

$$= 1 - p(0.99 \text{ mm} \leq X \leq 1.03 \text{ mm})$$

$$= 1 - p(0.99 \text{ mm} \leq X \leq 0.99 \text{ mm} + 0.04 \text{ mm})$$

$$= 1 - 10 \text{ mm}^{-1} \cdot 0.04 \text{ mm}$$

$$= 1 - 0.4$$

$$= 0.6.$$

Finally, let's consider the function that gives you the probability that X is smaller than a value x. Mathematically this is

$$F(x) = p(X \leq x).$$

This function has a special name. It is called the *cumulative distribution function* or CDF. We will examine its properties in detail as well in a later section. But for now, let's see how it looks like for our particular example. Pick an x smaller than 0.95 mm. Then, we should have $F(x) = 0$ because X cannot take values that are so small. What about an x greater than 1.05 mm. Then, we should have $F(x) = 1$ because X takes for sure values that are smaller than 1.05 m. Now, for an x between $0.95\,\text{mm}$ and $1.05\,\text{mm}$, we have

$$F(x) = p(X \leq x)$$
$$= p(0.95\text{ mm} \leq X \leq 0.95\text{ mm} + (x - 0.95\text{ mm}))$$
$$= 10\text{ mm}^{-1} \cdot (x - 0.95\text{ mm}).$$

Putting everything together, we find

$$F(x) = p(X \leq x)$$
$$= \begin{cases} 0, & \text{if } x < 0.95\text{ mm}, \\ 10\text{ mm}^{-1} \cdot (x - 0.95\text{ mm}), & \text{if } 0.95\text{ mm} \leq x \leq 1.05\text{ mm}, \\ 1, & \text{otherwise.} \end{cases}$$

You can use the CDF to find probabilities about X. Of course, $F(x)$ is the probability $p(X \leq x)$. But you can also use it to find the probability that X is greater than x. Here is how

$$p(X > 1.02\text{ mm}) = 1 - p(X \leq 1.02\text{ mm})$$
$$= 1 - F(1.02\text{ mm})$$
$$= 1 - 10\text{ mm}^{-1} \cdot (1.02\text{ mm} - 0.95\text{ mm})$$
$$= 1 - 10\text{ mm}^{-1} \cdot 0.07\text{ mm}$$
$$= 1 - 0.7$$
$$= 0.3.$$

10.2   The Cumulative Distribution Function

Take a continuous random variable X. The *cumulative distribution function* (CDF) $F_X(x)$ gives you the probability that X is smaller than x, i.e., the definition is

$$F_X(x) := p(X \leq x).$$

Note that the X is a subscript here. This is because each random variable has its own CDF. Now to save me some typing, I am going to write $F(x)$ instead of $F_X(x)$. We only have one random variable anyway.

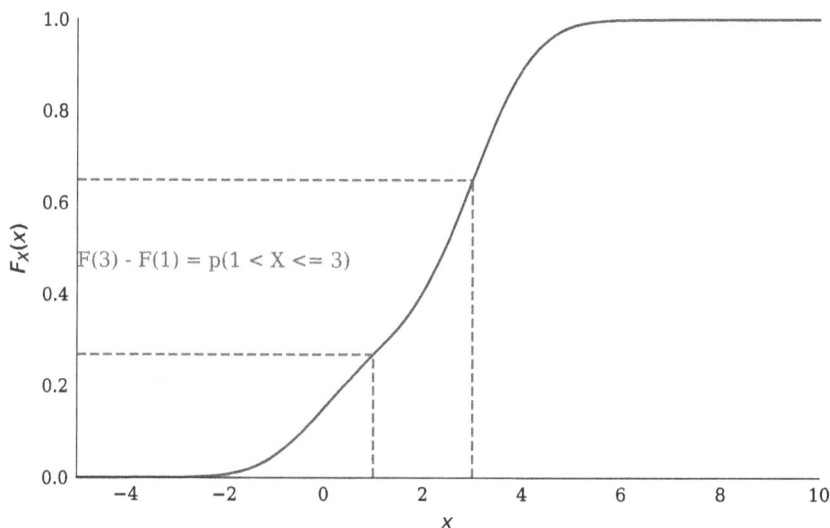

Figure 10.1. The CDF of a typical continuous random variable looks like this. It is an increasing function. It converges to zero for very small values. It converges to one for very large values. The probability of the random variable being within a given interval is the difference of CDF values at the endpoints of that interval. In this figure, $p(1 < X \le 3) = F(3) - F(1)$.

There are certain properties that the CDF satisfies. We visualize the properties in Figure 10.1. It is very instructive to go through the proof of these properties. The remarkable thing is that the proofs only use the basic probability rules.

10.2.1 *The CDF is an increasing function*

This property makes a lot of intuitive sense if you think about what the CDF means:

- The CDF at x is the probability that the random variable X is smaller than x.
- If you increase x, then the probability that X is smaller than x should increase.

This is pretty much it. But let's actually prove it rigorously. Take two values x_1 and x_2 such that $x_1 < x_2$. We need to show that

$F(x_1) \leq F(x_2)$. Here it is

$$
\begin{aligned}
F(x_2) &= p(X \leq x_2) \text{ (definition)} \\
&= p(X \leq x_1 \text{ or } x_1 < X \leq x_2) \text{ (logic)} \\
&= p(X \leq x_1) + p(x_1 < X \leq x_2) \text{ (sum rule)} \\
&= F(x_1) + p(x_1 < X \leq x_2) \text{ (definition)} \\
&\geq F(x_1),
\end{aligned}
$$

because $p(x_1 < X \leq x_2) \geq 0$. This concludes the proof.

10.2.2 *The CDF goes to zero for very small input values*

Mathematically, this is written as

$$
F(-\infty) = \lim_{x \to -\infty} F(x) = 0.
$$

Again, this is easy to understand:

- The CDF at x is the probability that the random variable X is smaller than x.
- Make the x very very small, smaller than any of the values X can take, and you get zero probability.

10.2.3 *The CDF goes to one for very large input values*

Mathematically, this is written as

$$
F(+\infty) = \lim_{x \to +\infty} F(x) = 1.
$$

Use your intuition:

- The CDF at x is the probability that the random variable X is smaller than x.
- Make the x very very large, larger than any of the values X can take, and you get probability one.

10.2.4   *Probability of values in an interval*

This property says that the probability of the random variable X taking values inside an interval $[a, b]$ is given by the difference of the values of CDF at b and a. Mathematically,

$$p(a < X \leq b) = F(b) - F(a).$$

This is obviously extremely useful because it tells you how to find any interval probability if you know the CDF. The proof is not that trivial. But it only uses the common rules. Let's do it:

$$p(a < X \leq b) = 1 - p(X \leq a \text{ or } X > b) \text{ (obvious rule)}$$
$$= 1 - [p(X \leq a) + p(X > b)] \text{ (sum rule)}$$
$$= 1 - [p(X \leq a) + 1 - p(X \leq b)] \text{ (obvious rule)}$$
$$= p(X \leq b) - p(X \leq a) \text{ (definition)}$$
$$= F(b) - F(a).$$

A corollary of this property is that if $F(x)$ is a continuous function for all x, then the probability that X takes exactly a given value is always zero. Here is how you can see this. Take the probability that X is between x and $x + \Delta x$ for some small Δx. It is

$$p(x < X \leq x + \Delta x) = F(x + \Delta x) - F(x).$$

Now, take $\Delta x \to 0$. The left-hand side goes to $p(X = x)$. The right-hand side goes to zero if $F(x)$ is continuous at x. Thus,

$$p(X = x) = 0,$$

if $F(x)$ is continuous at x.

All of the continuous random variables we will consider in this class satisfy this property.

10.3   The Probability Density Function

Take a continuous random variable X. The *probability density function* (PDF) $f_X(x)$ gives you the probability per unit of x that X is in a very small region around x. Intuitively, the definition is

$$f_X(x) \approx \frac{p(x < X \le x + \Delta x)}{\Delta x},$$

for Δx very, very small. In other words, $f_X(x)$ is the probability that X is between x and $x + \Delta x$ divided by Δx. It is exactly because you divide by Δx that this is a *probability density* and not just a probability.

As you have already realized, I do not like writing $f_X(x)$. So, if there is no ambiguity, I will write instead:

$$p(x) \equiv f_X(x).$$

Note 10.1. Again, this is not the precise mathematical definition of the PDF of a random variable, but it is good enough for our purposes.

There are some properties that the PDF satisfies which you absolutely need to remember. We visualize the most important properties in Figure 10.2.

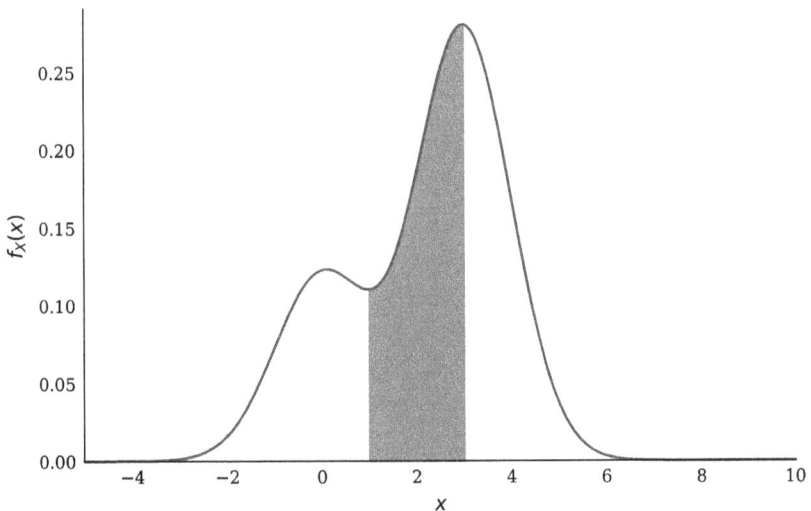

Figure 10.2. The PDF of a typical continuous random variable looks like this. It is a nonnegative function. The area between the function and the x-axis is one. The probability of the random variable being within a given interval is given by the integral of the PDF over that interval. In this figure, $p(1 < X \le 3) = \int_1^3 p(x)dx$ is the shaded area.

10.3.1 *The PDF is nonnegative*

Because $p(x)$ is a *probability* density and probabilities are nonnegative, we get that

$$p(x) \geq 0.$$

The set of x's for which $p(x)$ is strictly positive is known as the *support* of the PDF. Obviously, a random variable cannot take any values outside its support.

10.3.2 *The PDF is the derivative of the CDF*

Let $F(x)$ be the CDF of X. Then we have

$$p(x) = \frac{dF(x)}{dx}.$$

This is interesting! Before proving it, let's check the units. $F(x)$ has units of probability. If we divide by dx we get probability density. This looks promising. Let's do the proof.

We have

$$p(x) \approx \frac{p(x < X \leq x + \Delta x)}{\Delta x} \text{ (definition)}$$

$$= \frac{F(x + \Delta x) - F(x)}{\Delta x} \text{ (CDF Property 4)}.$$

Now, the result follows by taking the limit $\Delta x \to 0$.

As a sanity check, let's look back at our example. Does the property hold for what we found there? Our CDF was

$$F(x) = \begin{cases} 0, & \text{if } x < 0.95 \text{ mm,} \\ 10 \text{ mm}^{-1} \cdot (x - 0.95 \text{ mm}), & \text{if } 0.95 \text{ mm} \leq x \ 1.05 \text{ mm,} \\ 1, & \text{otherwise.} \end{cases}$$

Taking the derivative with respect to x yields:

$$p(x) = \begin{cases} 0, & \text{if } x < 0.95 \text{ mm,} \\ 10 \text{ mm}^{-1}, & \text{if } 0.95 \text{ mm} \leq x \ 1.05 \text{ mm,} \\ 0, & \text{otherwise,} \end{cases}$$

which agrees with the constant density we found in our example.

10.3.3   The CDF is the integral of the PDF

We have:

$$F(x) = \int_{-\infty}^{x} p(\tilde{x})d\tilde{x}.$$

This follows directly from PDF Property 2, if we employ the fundamental theorem of calculus.

10.3.4   Probability the random variable takes values in a set is the integral of the PDF over the set

The probability that X takes values between a and b for $a < b$ is given by

$$p(a < X \le b) = \int_{a}^{b} p(x)dx.$$

See Figure 10.2 for a visualization. The probability is the area between the PDF, the x-axis, and the endpoints of the interval $[a, b]$.

This is obviously extremely useful because if you have the PDF it allows you to easily calculate probabilities by doing integrals! Let's prove it:

$$p(a < X \le b) = F(b) - F(a) \text{ (CDF Property 4)}$$

$$= \int_{a}^{b} \frac{dF(x)}{dx}dx \text{ (fundamental theorem of calculus)}$$

$$= \int_{a}^{b} p(x)dx \text{ (PDF Property 2).}$$

This property can be generalize. If A is any set of possible values of X, then:

$$p(X \in A) = \int_{A} p(x)dx.$$

10.3.5   The PDF integrates to one

The property is

$$\int_{-\infty}^{+\infty} p(x)dx = 1.$$

In words, the total area between the PDF curve and the x-axis is one. Proving this property is obvious. You just have to note that:

$$\int_{-\infty}^{+\infty} p(x)dx = p(-\infty < X \leq +\infty) = 1,$$

because X must take some value.

It is a very common mistake to think that the PDF has to be smaller than one. But the PDF is a *probability density* not a *probability*. So, it can take values greater than one. In our example the probability density was 10 mm^{-1}.

10.4 The Uniform Distribution over the Unit Interval: Colab

In Section 10.1, we had a random variable for which all values between a given interval were equally probable. This is a very common situation covered by the so-called *uniform distribution*.

Let's start with the simplest case: a random variable taking values between 0 and 1 with constant probability density. We write

$$X \sim U([0,1]),$$

and we read

> The random variable X follows a uniform distribution taking values in $[0, 1]$.

You can refer to X as a uniform random variable. The probability density of a uniform random variable is constant in $[0, 1]$ and zero outside it. We have

$$p(x) := \begin{cases} c, & 0 \leq x \leq 1, \\ 0, & \text{otherwise.} \end{cases}$$

What should the constant c be? Just like in Section 10.1, you can find it by ensuring the PDF integrates to one (see Section 10.3.5):

$$\int_0^1 p(x)dx = 1 \Rightarrow c = 1.$$

So, the PDF is

$$p(x) := \begin{cases} 1, & 0 \le x \le 1, \\ 0, & \text{otherwise.} \end{cases}$$

To find the CDF, we can use Section 10.3.3:

$$F(x) = p(X \le x) = \int_0^x p(\tilde{x})d\tilde{x} = \int_0^x d\tilde{x} = x.$$

Obviously, we have $F(x) = 0$ for $x < 0$ and $F(x) = 1$ for $x > 1$.

Using this result, we can find the probability that X takes values in $[a, b]$ for $a < b$ in $[0, 1]$. It is

$$p(a \le X \le b) = F(b) - F(a) = b - a.$$

Let me know show you how you can make a uniform random variable using `scipy.stats`:

```
import scipy.stats as st
X = st.uniform()
```

Here are 10 samples:

```
X.rvs(size=10)
```

A possible output is

```
array([0.46421665, 0.94256114, 0.22811718, 0.95630123, 0.82063789,
       0.60783435, 0.3207602 , 0.1557496 , 0.20660041, 0.73482952])
```

You can evaluate the PDF at a given point:

```
X.pdf(0.5)
```

The output is

```
1.0
```

Another example:

```
X.pdf(-0.1)
```

The output is

```
0.0
```

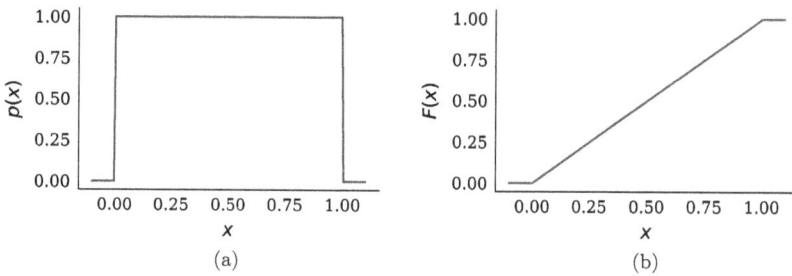

Figure 10.3. (a) The probability density function (PDF) and (b) cumulative distribution function (CDF) of a uniform random variable.

Here is how you can plot the PDF:

```
fig, ax = plt.subplots()
xs = np.linspace(-0.1, 1.1, 200)
ax.plot(xs, X.pdf(xs))
ax.set_xlabel('$x$')
ax.set_ylabel('$p(x)$')
```

The result is shown in Figure 10.3.

Finally, let's find the probability that X is between two numbers a and b. For the uniform it is, of course, trivial, but let's see how it is done using the scipy functionality:

```
a = 0.2
b = 0.8
prob_X_is_in_ab = X.cdf(b) - X.cdf(a)
print(f'p({a:1.2f} <= X <= {b:1.2f}) = {prob_X_is_in_ab:1.2f}')
```

The output is:

```
p(-1.00 <= X <= 0.30) = 0.30
```

10.5 The Uniform Distribution over an Arbitrary Interval

The uniform distribution can also be defined over an arbitrary interval $[a, b]$. We write

$$X \sim U([a, b]).$$

We read

The random variable X follows a uniform distribution on $[a, b]$.

The PDF of this random variable is

$$p(x) = \begin{cases} c, & x \in [a, b], \\ 0, & \text{otherwise,} \end{cases}$$

where c is a positive constant to be determined by the normalization condition:

$$\int_{-\infty}^{+\infty} p(x)dx = 1.$$

This gives

$$1 = \int_{-\infty}^{+\infty} p(x)dx = \int_a^b cdx = c\int_a^b dx = c(b-a).$$

From this, we get

$$c = \frac{1}{b-a},$$

and we can now write:

$$p(x) = \begin{cases} \frac{1}{b-a}, & x \in [a, b], \\ 0, & \text{otherwise,} \end{cases}$$

From the PDF, we can now find the CDF for $x \in [a, b]$:

$$F(x) = p(X \le x) = \int_{-\infty}^x p(\tilde{x})d\tilde{x} = \int_a^x \frac{1}{b-a}d\tilde{x}$$

$$= \frac{1}{b-a}\int_a^x d\tilde{x} = \frac{x-a}{b-a}.$$

Let's instantiate using `scipy.stats`:

```
a = -2.0
b = 5.0
X = st.uniform(loc=a, scale=(b-a))
```

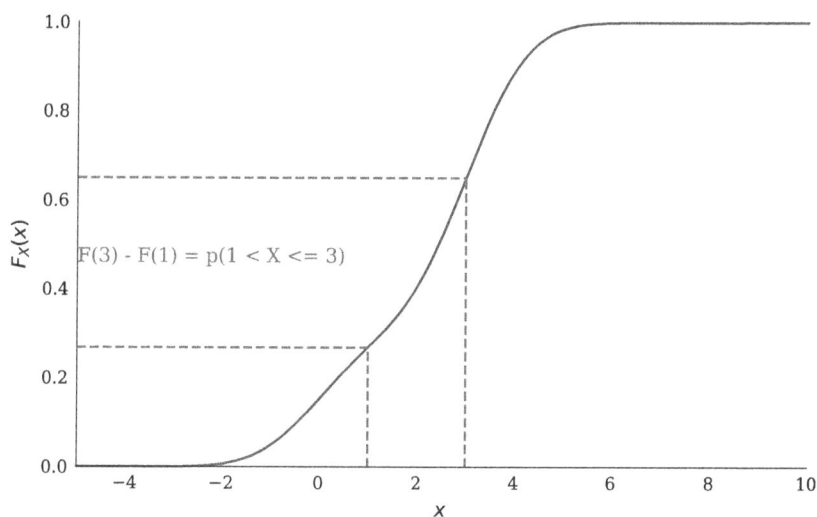

Figure 10.4. The probability density function (PDF) and cumulative distribution function (CDF) of a uniform random variable over the interval $[-2, 5]$.

Here are 10 samples:

```
X.rvs(size=10)
```

A possible output is

```
array([-1.4127092 ,  0.68758926,  3.92169569,  3.45911515,
 ↪   4.25865888,
         3.6047341 , -0.29694871,  0.13752324,  0.00893064,
         ↪   -1.86788956])
```

Figure 10.4 shows the PDF and CDF of this random variable.

Exercises

1. Which of the following experiments can be modeled with a continuous random variable (True/False):

   (a) Drawing two cards at random from a well-shuffled deck.
   (b) Weighing a piece of steel.
   (c) The radius of a single lamination of the stator of an electric machine punched out of a sheet of steel.
   (d) Gauging the pressure of a gas.

    (e) Counting the number of particles hitting a Geiger counter probe in one second.

2. You have a random variable X that measures the mass of an object. What are the units of its PDF $p(x)$?

3. Plot the PDF and CDF of a random variable with the uniform distribution over the interval $[1, 10]$.

4. **An alternative way to construct the generalized uniform distribution** $U([a, b])$. Let Z be a standard uniform random variable:

$$Z \sim U([0, 1]).$$

Define the random variable:

$$X = a + (b - a)Z.$$

(a) Show that:

$$X \sim U([a, b]).$$

Hint: Prove that the CDF of X is $F_X(x) = p(X \leq x) = \frac{x-a}{b-a}$.

(b) The function `numpy.random.rand` gives you uniform random samples in $[0, 1]$. Take 1,000 such samples and transform them to uniform samples in $[-1, 5]$. *Hint*: Fill in the missing code below.

```
a = -1
b = 5
z = np.random.rand(1000)
x = # Your code here
```

Test if you are getting the right answer by doing the histogram of your samples (it should be almost flat between -1 and 5):

5. **The Exponential distribution.** The Exponential distribution models the probability distribution of the time between events which occur continuously and independently at a constant rate. Examples of such a situation are as follows:

- The time between phone calls in a help center.
- The time between the arrival of cars at a toll station.
- The time between the arrival of customers.

- The time between two earthquakes.
- The time between two micrometeoroid impacts on a Moon research base.
- The time between faults in a mechanical system. However, this is a gross approximation because the rate of faults in a mechanical system increases with time, it is not constant.

We write

$$T \sim \text{Exp}(\lambda),$$

and we read

> The random variable T follows an Exponential distribution with rate parameter λ.

The rate parameter λ is positive and has units of inverse time. You can think of λ as the number of events per unit of time. The CDF of the Exponential is

$$F(t) = \begin{cases} 0, & t < 0, \\ 1 - e^{-\lambda t}, & t \geq 0. \end{cases}$$

(a) Prove mathematically that the PDF of the random variable T is:

$$p(t) = \lambda e^{-\lambda t}.$$

Hint: Use one of the properties of the PDF.

(b) Micrometeoroids, tiny particles of rock and dust, make space exploration very challenging. Even though the mass of these projectiles is very small (less than 1 gram), they move with a very high velocity (of the order of 10 km per second) and thus they will be gradually degrading the protective layers of any deep space habitat. For a Moon base with an area of 1000 squared meters, the rate of micrometeoroid impacts is about:

$$\lambda = 3 \times 10^{-6} \text{ s}^{-1}.$$

Read the `scipy.stats.expon` documentation and make an Exponential random variable T with this rate:

```
T = st.expon(scale=1/lambda)
```

(c) Take 1000 samples from the random variable you just constructed and draw their histogram.

(d) Plot the CDF of T.

(e) Plot the PDF of T.

(f) Find the probability that we have at least one micrometeoroid impact within a day. *Hint*: Remember that the units of T are seconds.

Chapter 11

Expectations, Variances, and Their Properties

In this chapter, I introduce the expectation and variance, statistical concepts that summarize the distribution of a random variable. I discuss their intuitive meaning and demonstrate some of their properties.

11.1 Expectation of Discrete Random Variables: Colab

Let X be a discrete random variable taking values $0, 1, 2, \ldots$. The *expectation* of X is defined by:

$$\mathbf{E}[X] = \sum_x x p(x),$$

where $p(x)$ is the PMF and the summation is over all discrete values of X. The expectation of a random variable is also known as the *mean* of the random variable.

You can think of the expectation as the value one should *expect* to get. However, take this interpretation with a grain of salt because the expectation may not even be a value that the random variable can take. Let's look at some examples.

11.1.1   *Expectation of a Bernoulli random variable*

Take a Bernoulli random variable (see Section 9.3):

$$X \sim \text{Bernoulli}(\theta).$$

Then:

$$\mathbf{E}[X] = \sum_x x p(x)$$
$$= 0 \cdot p(X = 0) + 1 \cdot p(X = 1)$$
$$= 0 \cdot (1 - \theta) + 1 \cdot \theta$$
$$= \theta.$$

And here is how we can do it using `scipy.stats`:

```
import scipy.stats as st
```

```
theta = 0.7
X = st.bernoulli(theta)
```

You can get the expectation like this:

```
print(f'E[X] = {X.expect():1.2f}')
```

The output is:

```
E[X] = 0.70
```

Figure 11.1 shows the PMF and the expectation of this Bernoulli random variable.

Notice that the expectation is not a value that the random variable can take. What is an intuitive way to think about the expectation? You can think of the random variable X as a rod of unit length with two masses attached to each end. The mass at the left end is $1 - \theta$ and the mass at the right end is θ. The expectation is the point on the rod where the rod balances, i.e., the center of mass.

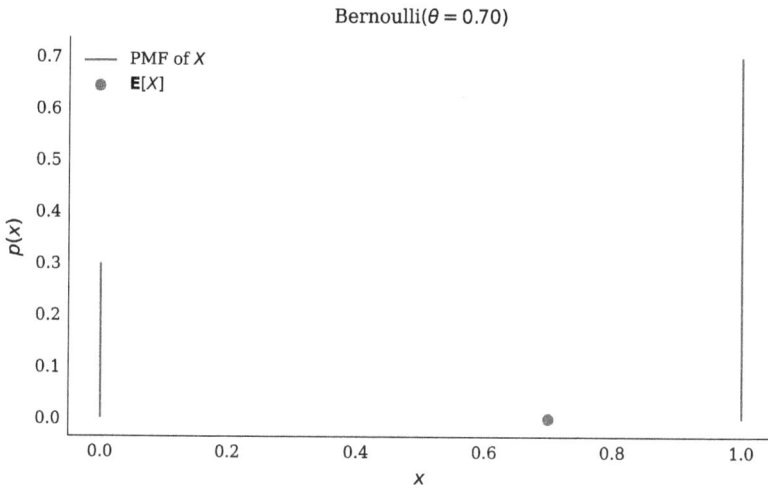

Figure 11.1. The PMF and the expectation of a Bernoulli random variable.

11.1.2 *Expectation of a categorical random variable*

Take a Categorical random variable (see Section 9.5):

$$X \sim \text{Categorical}(0.1, 0.3, 0.4, 0.2).$$

The expectation is:

$$\mathbf{E}[X] = \sum_x x p(x)$$

$$= 0 \cdot p(X = 0) + 1 \cdot p(X = 1) + 2 \cdot p(X = 2) + 3 \cdot p(X = 3)$$

$$= 0 \cdot 0.1 + 1 \cdot 0.3 + 2 \cdot 0.4 + 3 \cdot 0.2$$

$$= 1.7.$$

Here is how we can find it with Python:

```
import numpy as np
# The values X can take
xs = np.arange(4)
print('X values: ', xs)
# The probability for each value
```

```
ps = np.array([0.1, 0.3, 0.4, 0.2])
print('X probabilities: ', ps)
# And the expectation in a single line
E_X = np.sum(xs * ps)
print(f'E[X] = {E_X:1.2f}')
```

The output is:

```
X values:   [0 1 2 3]
X probabilities:   [0.1 0.3 0.4 0.2]
E[X] = 1.70
```

Alternatively, we could use `scipy.stats` to find the expectation:

```
X = st.rv_discrete(name='X', values=(xs, ps))
print(f'E[X] = {X.expect():1.2f}')
```

The output is:

```
E[X] = 1.70
```

Figure 11.2 shows the PMF and the expectation of this Categorical random variable. Again, you can think of the random variable X as a rod. The length of the rod is now 5 units and there is a mass at the beginning and at each unit step. The mass at the beginning is 0.1, the mass at the first unit step is 0.3, the mass at the second unit step is 0.4, and the mass at the third unit step is 0.2. The expectation is the point on the rod where the rod balances.

11.1.3   *Expectation of a binomial random variable*

Take a Binomial random variable (see Section 9.6):

$$X \sim \text{Binomial}(n, \theta).$$

The expectation is:

$$\mathbf{E}[X] = \sum_x x p(x)$$
$$= \sum_{k=0}^{n} k \binom{n}{k} \theta^k (1 - \theta)^{n-k}$$
$$= n\theta,$$

where we changed x for k since it is just a discrete dummy index.

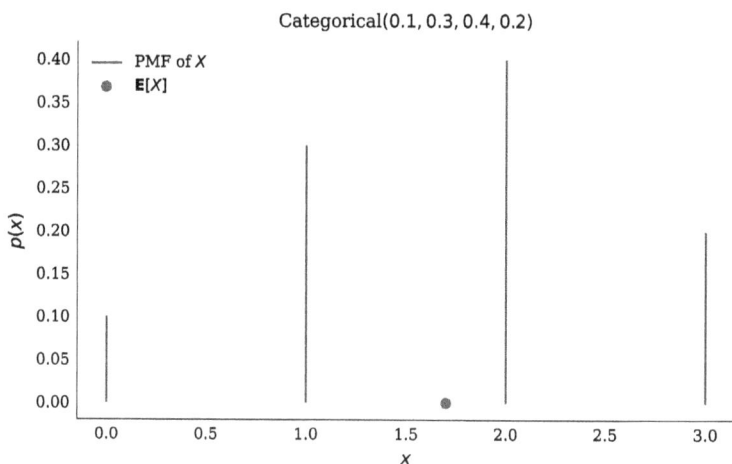

Figure 11.2. The PMF and the expectation of a Categorical random variable.

It is not trivial to prove this formula. The first step requires showing that for $k > 0$:

$$k\binom{n}{k} = n\binom{n-1}{k-1}.$$

You can prove it by expanding both sides and seeing that they are equal[1]:

$$k\frac{n!}{k!(n-k)!} \stackrel{?}{=} n\frac{(n-1)!}{(k-1)!(n-k)!}.$$

Meditate a bit on this and you will see it. Move the k from the left to the right. Then focus on the right side. Let both the n in the numerator and the k in the denominator be absorbed by the factorials.

For the second step, we just use the identity and the definition of the expectation:

$$\mathbf{E}[X] = \sum_{k=0}^{n} k\binom{n}{k}\theta^k(1-\theta)^{n-k} \qquad \text{(definition)}$$

$$= \sum_{k=1}^{n} k\binom{n}{k}\theta^k(1-\theta)^{n-k} \qquad (k = 0 \text{ does not contribute to the sum)}$$

---

[1] The $\stackrel{?}{=}$ symbol means that we are trying to check if the two sides are equal.

$$= \sum_{k=1}^{n} n \binom{n-1}{k-1} \theta^k (1-\theta)^{n-k} \qquad \text{(using the identity)}$$

$$= n \sum_{k=0}^{n} \binom{n-1}{k-1} \theta^k (1-\theta)^{n-k} \qquad \text{(taking } n \text{ out)}$$

$$= n \sum_{k'=0}^{n-1} \binom{n-1}{k'} \theta^{k'+1} (1-\theta)^{n-k'-1} \qquad (k' = k - 1)$$

$$= n\theta \sum_{k'=0}^{n-1} \binom{n-1}{k'} \theta^{k'} (1-\theta)^{n-1-k'} \qquad \text{(taking } \theta \text{ out)}$$

$$= n\theta \cdot 1,$$

where in the last step we used the fact that $\sum_{k'=0}^{n-1} \binom{n-1}{k'} \theta^{k'} (1-\theta)^{n-1-k'} = 1$ because it is the sum of the PMF of a Binomial random variable with parameters $n - 1$ and θ.

Here is how we can get the expectation with `scipy.stats`:

```
n = 5
theta = 0.6
X = st.binom(n, theta)
print(f'E[X] = {X.expect():1.2f}')
print(f'Compare to n * theta = {n * theta:1.2f}')
```

The output is:

```
E[X] = 3.00
Compare to n * theta = 3.00
```

Figure 11.3 shows the PMF and the expectation of this Binomial random variable.

11.1.4 *Expectation of a Poisson random variable*

Take a Poisson random variable (see Section 9.7):

$$X \sim \text{Poisson}(\lambda).$$

Binomial($n = 5, \theta = 0.60$)

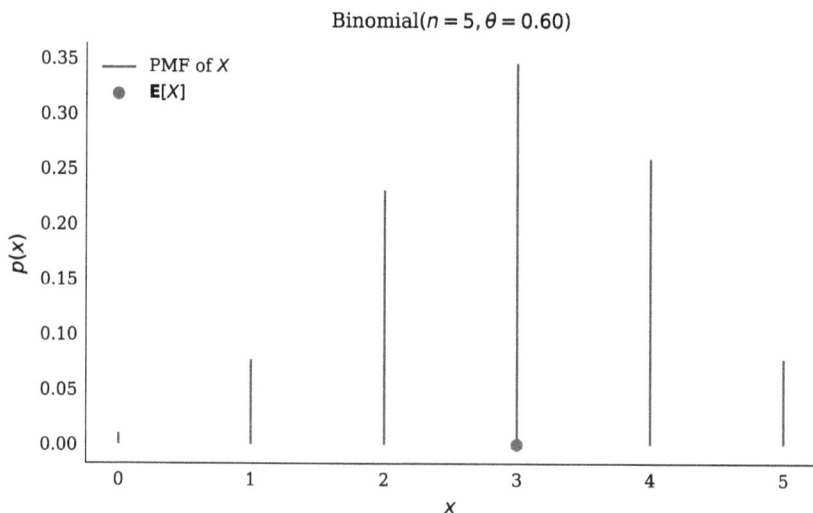

Figure 11.3. The PMF and the expectation of a Binomial random variable.

The expectation is[2]:

$$\mathbf{E}[X] = \sum_{k=0}^{\infty} k \frac{\lambda^k e^{-\lambda}}{k!} \qquad \text{(definition)}$$

$$= e^{-\lambda} \sum_{k=0}^{\infty} k \frac{\lambda^k}{k!} \qquad \text{(take exponential out of the sum)}$$

$$= e^{-\lambda} \sum_{k=1}^{\infty} k \frac{\lambda^k}{k!} \qquad \text{(the zero term does not contribute to the sum)}$$

$$= e^{-\lambda} \sum_{k=1}^{\infty} \frac{\lambda^k}{(k-1)!} \qquad \text{(cancel out terms)}$$

$$= e^{-\lambda} \lambda \sum_{k=1}^{\infty} \frac{\lambda^{k-1}}{(k-1)!} \qquad \text{(take one } \lambda \text{ out of the sum)}$$

---

[2]Recall that $e^{\lambda} = \sum_{k=0}^{\infty} \frac{\lambda^k}{k!}$.

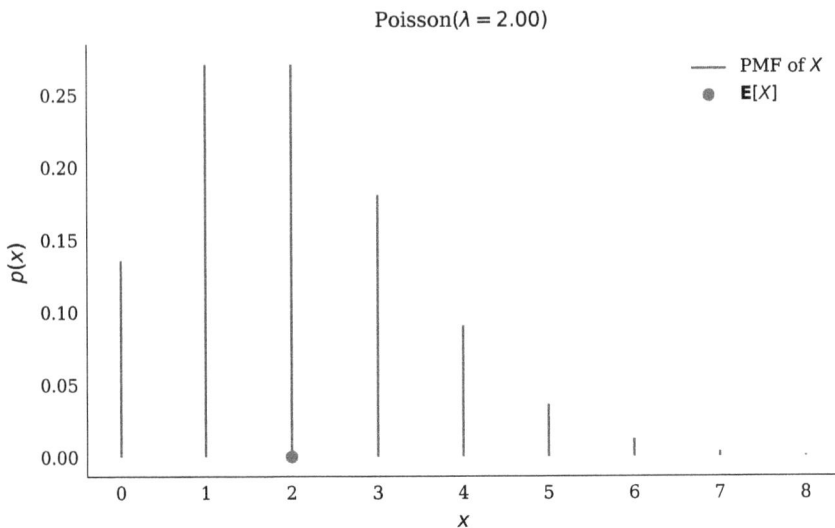

Figure 11.4. The PMF and the expectation of a Poisson random variable.

$$= e^{-\lambda}\lambda \sum_{k'=0}^{\infty} \frac{\lambda^{k'}}{k'!} \qquad \text{(rename the index)}$$

$$= e^{-\lambda}\lambda e^{\lambda} \qquad \text{(use exponential identity)}$$

$$= \lambda.$$

Let's also do it in `scipy.stats`:

```
lam = 2.0
X = st.poisson(lam)
print(f'E[X] = {X.expect():1.2f}')
```

The output is:

```
E[X] = 2.00
```

Figure 11.4 shows the PMF and the expectation of this Poisson random variable.

11.2 Expectation of a Continuous Random Variable

Let X be a continuous random variable taking real values. The expectation of X is defined by:

$$\mathbf{E}[X] = \int_{-\infty}^{+\infty} xp(x),$$

where $p(x)$ is the PDF of X.

11.2.1 *The expectation of a uniform random variable*

Take a uniform random variable:

$$X \sim U([a, b]).$$

Its expectation is:

$$\mathbf{E}[X] = \int_{-\infty}^{+\infty} xp(x)dx$$

$$= \int_{a}^{b} x\frac{1}{b-a}dx, \text{ (PDF is zero outside the interval)}$$

$$= \frac{1}{b-a}\frac{x^2}{2}\Big|_{a}^{b}$$

$$= \frac{1}{b-a}\frac{b^2-a^2}{2}$$

$$= \frac{1}{b-a}\frac{(b-a)(b+a)}{2}$$

$$= \frac{a+b}{2},$$

which makes a lot of sense because it is the point between a and b.

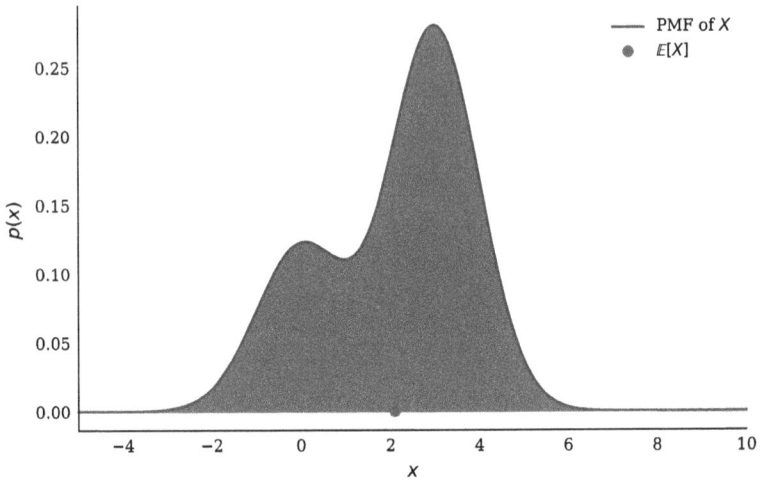

Figure 11.5. The expectation of a random variable is the x-coordinate of the centroid of the area between the x-axis and the PDF.

11.2.2 *Interpretation of the expectation of continuous random variables*

Mathematically, the expectation is the x-coordinate of the centroid of the area between the x-axis and the PDF, see Figure 11.5.

As in the discrete case, you can think of the expectation as the value one should "expect" to get. But, again, take this interpretation with a grain of salt... See Figure 11.6 for an example where the expectation is a very improbable value.

In Figure 11.6, the PDF of the random variable has two distinct peaks. Each peak of the PDF is called a mode. "Nice" PDF's have a single mode. They are called unimodal. For unimodal PDF's the interpretation that the expectation is the value you should expect to get makes a lot of sense. However, a PDF may be multi-modal. Then the expectation is not very useful.

11.3 Simplifying Our Notation about Expectations

We saw that for discrete random variables we need to write:

$$\mathbf{E}[X] = \sum_x xp(x),$$

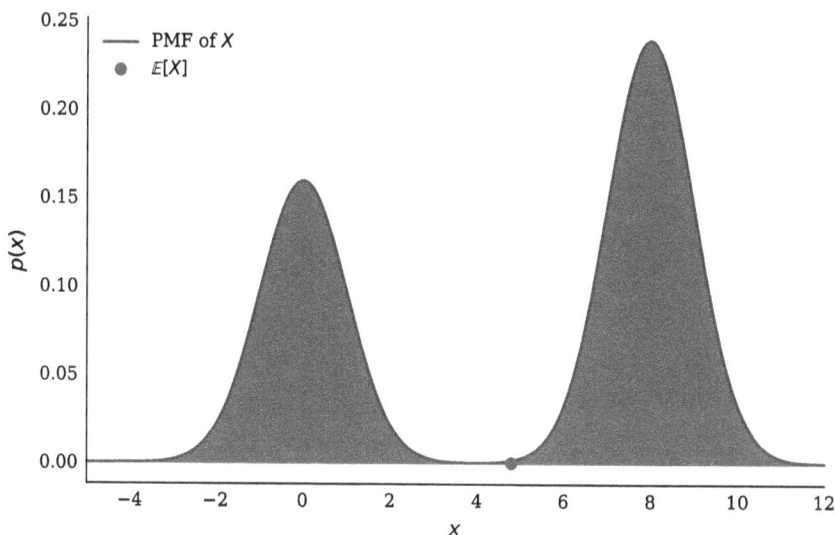

Figure 11.6. The expectation of a random variable does not have to be on a high probability region.

while for a continuous random variable we need to write:

$$\mathbf{E}[X] = \int_{-\infty}^{+\infty} x p(x).$$

Let's make the following convention. We are going to write:

$$\mathbf{E}[X] = \int x p(x),$$

no matter what the nature of X is. If it is discrete, we will sum over its values. If it is continuous, we will integrate over its values. This is the notation most commonly used in practice.

11.4   Properties of Expectations

There are a few very important properties that expectations satisfy. Let's introduce them one by one. In what follows X is a random variable, either discrete or continuous.

11.4.1   *The expectation of a scalar times a random variable*

Take any scalar λ. Then:

$$\mathbf{E}[\lambda X] = \lambda \mathbf{E}[X].$$

The proof is super easy. It follows directly from the properties of integrals and summations.

11.4.2   *The expectation of a function of a random variable*

Let $f(x)$ be any function defined over the support of the random variable X. Then:

$$Z = f(X),$$

is a new random variable. It is just a random variable that is induced by transforming the values generated by X through the function $f(x)$. Then, the expectation of Z is:

$$\mathbf{E}[Z] \equiv \mathbf{E}[f(X)] = \int f(x)p(x)dx.$$

We will not prove this property. The proof is non-trivial and requires expressing the PDF of $Z = f(X)$ in terms of the PDF of X. Because people tend to take this property for granted, it is known as the *law of the unconscious statistician*.

This formula is, of course, extremely convenient for calculating expectations of functions of random variables. Let's see an example. Assume that X is a uniform:

$$X \sim U([0, 1]).$$

Then:

$$\mathbf{E}[X^2] = \int x^2 p(x)dx$$

$$= \int_0^1 x^2 dx$$

$$= \left.\frac{x^3}{3}\right|_0^1$$

$$= \frac{1}{3}.$$

Similarly,

$$\mathbf{E}[e^X] = \int_0^1 e^x dx = e.$$

And so on and so forth.

I have to say this because I have seen many students do this. You may get the urge to write:

$$\mathbf{E}[f(X)] = f(\mathbf{E}[X]). \text{ (WRONG)}.$$

Please don't. This formula does not hold unless $f(x)$ is an affine function (i.e., $f(x) = ax + b$). Things are way more complicated than this. For example, for convex functions you have that[3]:

$$f(\mathbf{E}[X]) \le \mathbf{E}[f(X)] \ (f \text{convex}).$$

For concave functions, you have the opposite inequality. For arbitrary functions, it could go in any direction.

11.4.3   *The expectation of a random variable plus a scalar*

This property is a direct corollary of the second property, but it is worth spelling it out on its own. Take a scalar λ. Then:

$$\mathbf{E}[X + \lambda] = \mathbf{E}[X] + \lambda.$$

The property makes a lot of intuitive sense. Let's actually do the proof of this one. We have:

$$\mathbf{E}[X + \lambda] = \int (x + \lambda) p(x) dx$$

(property 2)

---

[3]This is known as Jensen's inequality.

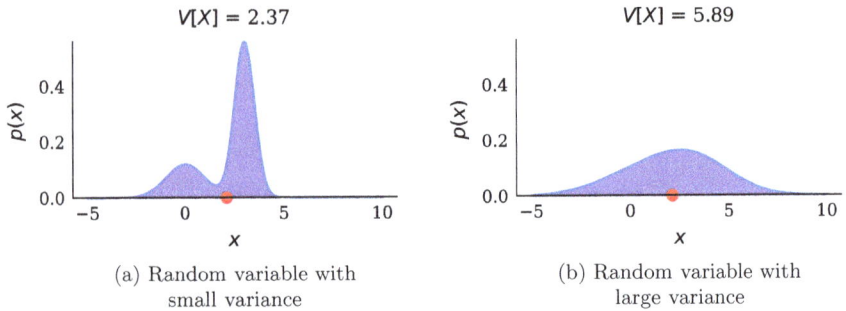

(a) Random variable with small variance

(b) Random variable with large variance

Figure 11.7. Comparison of the variance of two arbitrary random variables. We show the PDF of the random variables (blue shaded area) and the expectation (red dot). The random variable on the right has six times the variance of the one on the left.

$$= \int xp(x)dx + \lambda \int p(x)dx$$

(integral properties)

$$= \mathbf{E}[X] + \lambda$$

(PDF integrates to one).

11.5   Variance of a Random Variable

Take a random variable X (either continuous or discrete). Assume that the expectation is some number:

$$\mu = \mathbf{E}[X].$$

The variance $\mathbf{V}[X]$ of X is defined as the expectation of the squared difference of X from the expected value (or the mean) μ, i.e.,

$$\mathbf{V}[X] := \mathbf{E}\left[(X - \mu)^2\right].$$

You can think of the variance as the measure of the spread of the random variable around its expected value. See Figure 11.7 for a visualization.

11.6 Properties of Variance

The variance operator satisfies certain important properties. Let's go over the most important ones with some proofs. In what follows X is either a discrete or a continuous random variable and $\mu = \mathbf{E}[X]$ is its expected value.

11.6.1 *The variance of a random variable times a scalar is the square of the scalar times the variance of the random variable*

Let λ be any scalar value. Then:

$$\mathbf{V}[\lambda X] = \lambda^2 \mathbf{V}[X].$$

To prove this, we need to invoke the definition of the variance. First, notice from from the first property of expectations, we have:

$$\mathbf{E}[\lambda X] = \lambda \mathbf{E}[X] = \lambda \mu.$$

So, the variance of λX is by definition:

$$\begin{aligned}
\mathbf{V}[\lambda X] &= \mathbf{E}[(\lambda X - \lambda \mu)^2] \\
&= \mathbf{E}[\lambda^2 (X - \mu)^2] \\
&= \lambda^2 \mathbf{E}[(X - \mu)^2] \\
&= \lambda^2 \mathbf{V}[X].
\end{aligned}$$

11.6.2 *The variance of a random variable plus a constant is the variance of the random variable*

Let λ be any number. Then:

$$\mathbf{V}[X + \lambda] = \mathbf{V}[X].$$

And here is the proof. The expectation of $X + \lambda$ is:

$$\mathbf{E}[X + \lambda] = \mathbf{E}[X] + \lambda = \mu + \lambda.$$

Then, by definition, the variance of $X + \lambda$ is:

$$\mathbf{V}[X + \lambda] = \mathbf{E}[(X + \lambda - \mu - \lambda)^2]$$
$$= \mathbf{E}[(X - \mu)^2]$$
$$= \mathbf{V}[X].$$

11.6.3   *The variance is the expectation of the square minus the expectation of a random variable*

Mathematically, we have:

$$\mathbf{V}[X] = \mathbf{E}[X^2] - \mu^2.$$

Going through the proof of this property can help you develop your understanding of mathematics. Here we go:

$$\mathbf{V}[X] = \mathbf{E}\left[(X - \mu)^2\right]$$
$$= \mathbf{E}\left[X^2 - 2\mu X + \mu^2\right] \qquad \text{(expanding the square)}$$
$$= \mathbf{E}[X^2] - \mathbf{E}[2\mu X] + \mu^2 \qquad \text{(expectation property)}$$
$$= \mathbf{E}[X^2] - 2\mu\mathbf{E}[X] + \mu^2 \qquad \text{(expectation property)}$$
$$= \mathbf{E}[X^2] - 2\mu \cdot \mu + \mu^2 \qquad \text{(definition of expectation)}$$
$$= \mathbf{E}[X^2] - 2\mu^2 + \mu^2$$
$$= \mathbf{E}[X^2] - \mu^2.$$

Now you may wonder why on Earth you need to know this property. Well, it turns out that this property offers one of the easiest ways to calculate the variance of random variables by hand. Let's look at some examples.

11.7   Examples of Variances of Random Variables: Colab

Let's revisit some of the distributions we encountered earlier and calculate their variances. We will do it both analytically, and using `scipy.stats`.

11.7.1   *Variance of a Bernoulli random variable*

Take a Bernoulli random variable (see Section 9.3):

$$X \sim \text{Bernoulli}(\theta).$$

We have already found that this expectation is:

$$\mu = \mathbf{E}[X] = \theta.$$

To find the variance, we are going to use the third property of variance from Section 11.6. For this, we need to find the expectation of the square:

$$\mathbf{E}[X^2] = \sum_x x^2 p(x)$$
$$= 0^2 p(X = 0) + 1^2 p(X = 1)$$
$$= \theta.$$

So, we have:

$$\mathbf{V}[X] = \mathbf{E}[X^2] - \mu^2 = \theta - \theta^2 = \theta(1 - \theta).$$

And here is how we can do it using `scipy.stats`:

```
theta = 0.7
X = st.bernoulli(theta)
print(f'V[X] = {X.var():1.2f}')
print(f'Compare to theta * (1 - theta) = {theta * (1 -
↪   theta):1.2f}')
```

The output is:

```
V[X] = 0.21
Compare to theta * (1 - theta) = 0.21
```

The standard deviation is just the square root of the variance:

$$\sigma = \sqrt{\theta(1 - \theta)}.$$

In `scipy.stats` you can get it by `X.std()`:

```
print(f'std of X = {X.std():1.2f}')
```

The output is:

```
std of X = 0.46
```

11.7.2 *Variance of a uniform random variable*

Take

$$X \sim U([a, b]).$$

Remember that the PDF is:

$$p(x) = \frac{1}{b - a},$$

when x is in $[a, b]$ and zero otherwise. We have already found the expectation and it was given by the mid-point between a and b:

$$\mu = \mathbf{E}[X] = \frac{a + b}{2}.$$

To find the variance, we first need to find the expectation of the square:

$$\mathbf{E}[X^2] = \int x^2 p(x) dx$$

$$= \int_a^b x^2 \frac{1}{b - a} dx$$

$$= \frac{1}{b - a} \frac{x^3}{3} \Big|_a^b$$

$$= \frac{1}{b - a} \frac{b^3 - a^3}{3}$$

$$= \frac{b^3 - a^3}{3(b - a)}.$$

You can simplify this even more, but we won't bother. Now, put everything together:

$$\mathbf{V}[X] = \mathbf{E}[X^2] - \mu^2$$

$$= \frac{b^3 - a^3}{3(b - a)} - \frac{(a + b)^2}{4}$$

$$= \frac{4 \cdot (b^3 - a^3) - 3(b - a)(a + b)^2}{12(b - a)}$$

$$= \frac{4b^3 - 4a^3 - 3(b - a)(a^2 + 2ab + b^2)}{12(b - a)}$$

$$= \frac{4b^3 - 4a^3 - 3a^2b - 6ab^2 - 3b^3 + 3a^3 + 6a^2b + 3ab^2}{12(b - a)}$$

$$= \frac{(b - a)^3}{12(b - a)}$$

$$= \frac{(b - a)^2}{12}.$$

If you remember your basic mechanics, this is the second area moment of inertia of a beam about its center of mass. This is not an accident. Mathematically, the variance and the second area moments are exactly the same integrals.

Let's do it also on `scipy.stats`:

```
a = 0
b = 5
X = st.uniform(a, b)
print(f'V[X] = {X.var():1.2f}')
print(f'Compare to theoretical answer = {(b - a) ** 2 / 12:1.2f}')
```

The output is:

```
V[X] = 2.08
Compare to theoretical answer = 2.08
```

11.7.3   *Variance of a categorical random variable*

Take a Categorical random variable (see Section 9.5):

$$X \sim \text{Categorical}(0.1, 0.3, 0.4, 0.2).$$

The expectation is:

$$\mu = \mathbf{E}[X] = 1.7.$$

Again, we are going to invoke the third property of variance. We need the expectation of the square:

$$\mathbf{E}[X^2] = \sum_x x^2 p(x)$$

$$= 0^2 \cdot p(X = 0) + 1^2 \cdot p(X = 1) + 2^2 \cdot p(X = 2)$$
$$+ 3^2 \cdot p(X = 3)$$
$$= 0 \cdot 0.1 + 1 \cdot 0.3 + 4 \cdot 0.4 + 9 \cdot 0.2$$
$$= 3.7.$$

So, we have:

$$\mathbf{V}[X] = \mathbf{E}[X^2] - \mu^2 = 3.7 - 1.7^2 = 0.81.$$

Here is how we can find it with Python:

```python
import numpy as np
# The values X can take
xs = np.arange(4)
print('X values: ', xs)
# The probability for each value
ps = np.array([0.1, 0.3, 0.4, 0.2])
print('X probabilities: ', ps)
# And the expectation in a single line
E_X = np.sum(xs * ps)
# The expectation of the square
E_X2 = np.sum(xs ** 2 * ps)
# The variance
V_X = E_X2 - E_X ** 2
print(f'V[X] = {V_X:1.2f}')
```

The output is:

```
X values:  [0 1 2 3]
X probabilities:  [0.1 0.3 0.4 0.2]
V[X] = 0.81
```

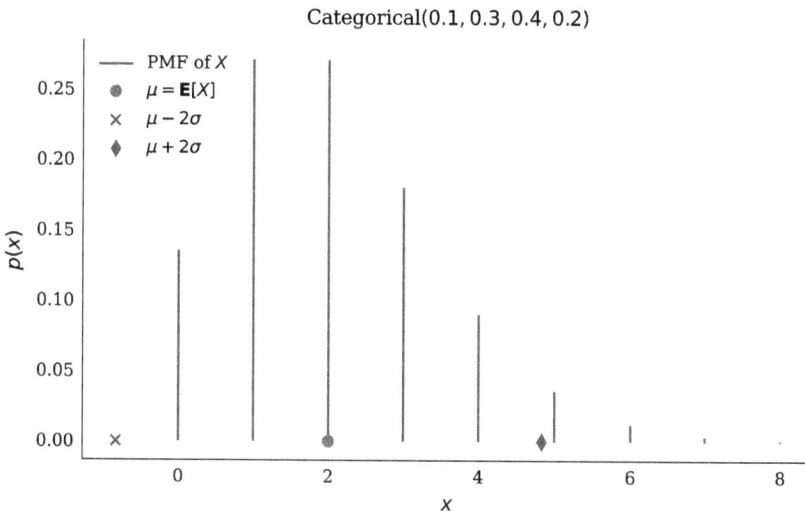

Figure 11.8. PMF of a Categorical random variable showing the expectation and two standard deviations around it. We see that going two standard deviations below and above the mean captures pretty much all the values.

Alternatively, we could use `scipy.stats`:

```
X = st.rv_discrete(name='X', values=(xs, ps))
print(f'V[X] = {X.var():1.2f}')
```

The output is:

```
V[X] = 0.81
```

The standard deviation is:

```
print(f'std of X = {X.std():1.2f}')
```

The output is:

```
std of X = 0.90
```

Let's now make a plot. I am going to plot the PMF of X and I am going to mark the position of the expected value along with:

- the expected value minus two standard deviations,
- the expected value plus two standard deviations.

Figure 11.8 shows this visualization.

11.7.4 *Variance of a binomial random variable*

Take a Binomial random variable (see Section 9.6):

$$X \sim \text{Binomial}(n, \theta).$$

The variance is given by:

$$\mathbf{V}[X] = n\theta(1 - \theta).$$

Let's prove it. We are going to use the standard identity:

$$\mathbf{V}[X] = \mathbf{E}[X^2] - \mu^2.$$

We already know that the expectation is:

$$\mu = \mathbf{E}[X] = n\theta.$$

Let's find the expectation of the square. We have:

$$\mathbf{E}[X^2] = \sum_{k=0}^{n} k^2 \binom{n}{k} \theta^k (1 - \theta)^{n-k} \qquad \text{(definition)}$$

$$= \sum_{k=1}^{n} k^2 \binom{n}{k} \theta^k (1 - \theta)^{n-k} \qquad (k = 0 \text{ does not contribute}$$
$$\text{to the sum)}$$

$$= \sum_{k=1}^{n} kn \binom{n-1}{k-1} \theta^k (1 - \theta)^{n-k} \qquad \text{(using the identity)}$$

$$= n\theta \sum_{k=1}^{n} k \binom{n-1}{k-1} \theta^{k-1} (1 - \theta)^{n-k} \qquad \text{(taking } n\theta \text{ out)}$$

$$= n\theta \sum_{k'=0}^{n-1} (k' + 1) \binom{n-1}{k'} \theta^{k'} (1 - \theta)^{n-1-k'} \qquad (k' = k - 1)$$

$$= n\theta \left[1 + (n - 1)\theta\right]$$

$$= n\theta(1 - \theta) + n^2\theta^2.$$

Going from the fifth to the sixth line, we used that

$$\sum_{k'=0}^{n-1} \binom{n-1}{k'} \theta^{k'} (1 - \theta)^{n-1-k'} = 1$$

and

$$\sum_{k'=0}^{n-1} k' \binom{n-1}{k'} \theta^{k'} (1-\theta)^{n-1-k'} = (n-1)\theta,$$

from the properties of the Binomial with parameters $n-1$ and θ. The desired result follows from plugging everything back into the identity for the variance.

Here is how we can get the variance of the Binomial with `scipy.stats`:

```
n = 5
theta = 0.6
X = st.binom(n, theta)
print(f'V[X] = {X.var():1.2f}')
print(f'Compare to n * theta * (1 - theta) = {n * theta * (1 -
↪    theta):1.2f}')
```

The output is:

```
V[X] = 1.20
Compare to n * theta * (1 - theta) = 1.20
```

Let's plot the same things we plotted for the categorical. Figure 11.9 shows the PMF with expectation and two standard deviations marked.

11.7.5 *Variance of a Poisson random variable*

Take a Poisson random variable (see Section 9.7):

$$X \sim \text{Poisson}(\lambda).$$

The variance is:

$$\mathbf{V}[X] = \lambda.$$

Wait a second!!! Didn't we say that the variance has the square units of X? If you paid attention, the expectation of X was also λ. How is it even possible? Well, it is because X has no units. It's just numbers counting events.

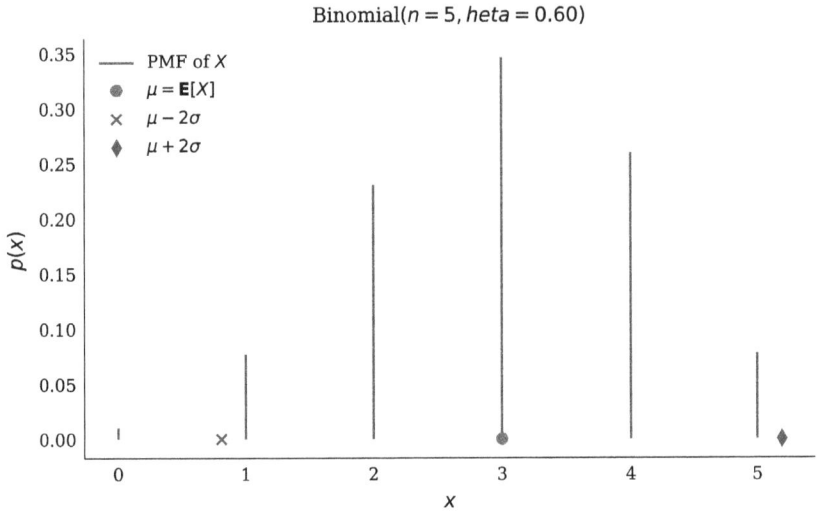

Figure 11.9. PMF of a Binomial random variable with $n = 5$ and $\theta = 0.60$, showing the expectation and two standard deviations around it.

Let's prove the variance formula. Let's work out the expectation of the square. We have:

$$\mathbf{E}[X^2] = \sum_{k=0}^{\infty} k^2 \frac{\lambda^k e^{-\lambda}}{k!} \qquad \text{(definition)}$$

$$= \sum_{k=1}^{\infty} k^2 \frac{\lambda^k e^{-\lambda}}{k!} \qquad \text{(drop zero term)}$$

$$= \lambda \sum_{k=1}^{\infty} k \frac{\lambda^{k-1} e^{-\lambda}}{(k-1)!} \qquad \text{(cancel one } k$$

$$\text{and take } \lambda \text{ out)}$$

$$= \lambda \sum_{k'=0}^{\infty} (k'+1) \frac{\lambda^{k'} e^{-\lambda}}{k'!} \qquad (k' = k - 1)$$

$$= \lambda \left(\sum_{k'=0}^{\infty} \frac{\lambda^{k'} e^{-\lambda}}{k'} + \sum_{k'=0}^{\infty} \frac{\lambda^{k'} e^{-\lambda}}{k'!} \right) \qquad \text{(split the sum)}$$

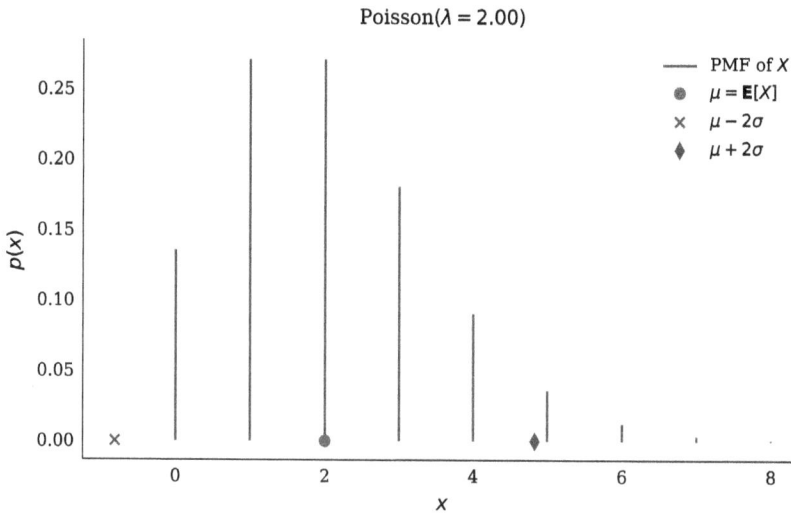

Figure 11.10. PMF of a Poisson random variable with $\lambda = 2.00$, showing the expectation and two standard deviations around it.

$$= \lambda\,(1 + \lambda) \qquad\qquad \text{(Poisson properties)}$$
$$= \lambda + \lambda^2.$$

Again, plugging everything back into the identity for the variance gives us the desired result.

Here is how we can get the variance of the Poisson with `scipy.stats`:

```
lam = 2.0
X = st.poisson(lam)
print(f'V[X] = {X.var():1.2f}')
print(f'Compare to lambda = {lam:1.2f}')
```

The output is:

```
V[X] = 2.00
Compare to lambda = 2.00
```

And let's visualize everything together like before. Figure 11.10 shows the PMF with expectation and two standard deviations marked.

Exercises

1. **Calculating expectations for discrete random variables**
   Consider the Categorical random variable $X \sim$ Categorical $(0.2, 0.4, 0.4)$ taking three discrete values $1, 2,$ and 3. Find the numerical answer for the following expressions. *Hint*: You can do it by hand or write some code. It is up to you.
   (a) $\mathbf{E}[X]$
   (b) $\mathbf{E}[X^2]$
   (c) $\mathbf{V}[X]$
   (d) $\mathbf{E}[e^X]$

2. **Calculate the expectation and variance of a continuous random variable**
   Take an exponential random variable with rate parameter λ:
   $$X \sim \text{Exp}(\lambda).$$
   The PDF of the exponential is:
   $$p(t) = \lambda e^{-\lambda t},$$
   for $t \geq 0$ and zero otherwise.
   (a) Prove mathematically that $\mathbf{E}[T] = \lambda^{-1}$. *Hint*: You need to do the integral $\int_0^\infty t p(t) dt$ using integration by parts.
   (b) Prove mathematically that $\mathbf{V}[T] = \lambda^{-2}$. *Hint*: Use integration by parts to find $\mathbf{E}[T^2] = \int_0^\infty t^2 p(t) dt$ and then use one of the properties of the variance.

3. **Properties of expectations**
   Let X be a random variable with expectation $\mathbf{E}[X] = 3$ and variance $\mathbf{V}[X] = 2$. Calculate the following expressions.
   (a) $\mathbf{E}[4X]$
   (b) $\mathbf{E}[X + 2]$
   (c) $\mathbf{E}[3X + 1]$
   (d) $\mathbf{V}[5X]$
   (e) $\mathbf{V}[X + 3]$
   (f) $\mathbf{V}[2X + 1]$
   (a) $\mathbf{E}[X^2]$

Chapter 12

The Normal Distribution, Quantiles, and Credible Intervals

In this chapter, I introduce the standard Normal distribution and its more general form, the Normal distribution. I explore quantiles and central credible intervals, statistical concepts used to summarize the distribution of a random variable. I demonstrate how to fit the parameters of the Normal distribution using the method of moments, and I show how to compare two Normal distributions using bootstrap.

12.1 The Standard Normal Distribution: Colab

The standard normal distribution is perhaps the most recognizable continuous distribution. We write:

$$Z \sim N(0, 1),$$

and we read it as:

> The random variable Z follows the standard Normal.

Now the 0 in $N(0, 1)$ is the expected value and the 1 is the variance.

12.1.1 *The PDF of the standard Normal*

We commonly use the function $\phi(z)$ to represent the PDF of the standard Normal. It is:

$$\phi(z) = \frac{1}{\sqrt{2\pi}} \exp\left\{-\frac{z^2}{2}\right\}.$$

This function is an exponential that has a z^2 inside it. The term $\frac{1}{\sqrt{2\pi}}$ is there so that the PDF is normalized, i.e.:

$$\int_{-\infty}^{+\infty} \phi(z)dz = 1.$$

Here is how you can make a standard Normal in "scipy.stats":

```
import scipy.stats as stats
Z = stats.norm()
```

Here are some samples:

```
Z.rvs(size=10)
```

A possible output is:

```
array([-0.07035267,  0.72710488, -1.424037  ,  1.43450515,
 ↪   -0.62655998,
         1.15299399, -0.40145701, -1.85201333, -0.4196824 ,
         ↪   -0.10409592])
```

The left panel of Figure 12.1 shows the PDF of the standard Normal.

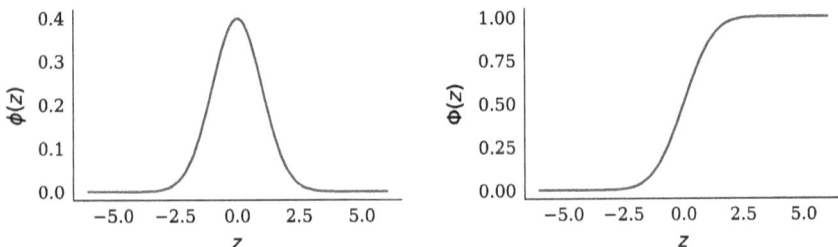

Figure 12.1.   The standard Normal distribution: PDF (left) and CDF (right).

Here are some important properties of the PDF of the standard Normal. First, $\phi(z)$ is positive for all z. Second, as z goes to $-\infty$ or $+\infty$, $\phi(z)$ goes to zero:

$$\lim_{z \to -\infty} \phi(z) = 0 \quad \text{and} \quad \lim_{z \to +\infty} \phi(z) = 0.$$

Third, $\phi(z)$ has a unique mode (maximum) at $z = 0$. In other words, $z = 0$ is the most probable point under this distribution. Fourth, $\phi(z)$ is symmetric about $z = 0$. Mathematically:

$$\phi(-z) = \phi(z).$$

Let's test some of these properties using `scipy.stats`. Try:

```
Z.pdf(np.inf)
```

You should get:

```
0.0
```

Similarly, for `Z.pdf(-np.inf)`. Try the symmetry property:

```
Z.pdf(-5) == Z.pdf(5)
```

You should get:

```
True
```

12.1.2 Expectation and variance of the standard Normal

The expectation of Z is zero:

$$\mathbf{E}[Z] = \int_{-\infty}^{+\infty} z\phi(z)dz = 0.$$

You can prove this quite easily by invoking the fact that $\phi(-z) = \phi(z)$. Here it is in `scipy.stats`:

```
Z.expect()
```

Output:

```
0.0
```

The variance of Z is one:

$$\mathbf{V}[Z] = \int_{-\infty}^{+\infty} z^2 \phi(z) dz = 1.$$

You need integration by parts to prove this. Let's do it.

$$\mathbf{V}[Z] = \int_{-\infty}^{+\infty} z^2 \phi(z) dz$$

$$= \frac{1}{\sqrt{2\pi}} \int_{-\infty}^{+\infty} z^2 \exp\left\{-\frac{z^2}{2}\right\} dz$$

$$= \frac{1}{\sqrt{2\pi}} \int_{-\infty}^{+\infty} (-z) \cdot \left[-z \exp\left\{-\frac{z^2}{2}\right\}\right] dz$$

$$= \frac{1}{\sqrt{2\pi}} \int_{-\infty}^{+\infty} z \cdot \frac{d}{dz}\left[\exp\left\{-\frac{z^2}{2}\right\}\right] dz$$

$$= -\frac{1}{\sqrt{2\pi}} \left\{\left[z \exp\left\{-\frac{z^2}{2}\right\}\right]_{-\infty}^{+\infty} - \int_{-\infty}^{+\infty} \exp\left\{-\frac{z^2}{2}\right\} dz\right\}$$

$$= -\frac{1}{\sqrt{2\pi}} \left(0 - \sqrt{2\pi}\right)$$

$$= 1.$$

Note that the standard deviation of Z is also 1 (Z does not have units). In `scipy.stats`, you can get the variance with:

```
Z.var()
```

You should get:

```
1.0
```

12.1.3 *The CDF of the standard Normal*

The CDF of Z gives you the probability that Z is smaller than a number z. It is common to use $\Phi(z)$ to denote the CDF. There is no closed form. But you can write this:

$$\Phi(z) = \int_{-\infty}^{z} \phi(z')dz',$$

where $\phi(z)$ is the PDF of Z. The right panel of Figure 12.1 visualizes the CDF of the standard Normal.

First, note that:

$$\Phi(0) = 0.5.$$

This follows very easily from the symmetry of the PDF $\phi(z)$ about zero. Remember that $\Phi(0)$ is the probability that Z is smaller than zero and because of symmetry that probability is exactly 50%. A point z with such a property (the probability that the random variable is smaller than it is 0.5) is called the *median* of the random variable.

Now here is a non-trivial property. Take any number z. Then, we have:

$$\Phi(-z) = 1 - \Phi(z).$$

Before attempting to prove this property, let's demonstrate it visually. Take z to be some positive number. Then $\Phi(-z)$ is the probability that Z is smaller than $-z$, or this area below the PDF:

$$\Phi(-z) = \int_{-\infty}^{-z} \phi(z')dz'.$$

On other hand, $1 - \Phi(z)$ is the following area:

$$1 - \Phi(z)$$

$$= \int_{-\infty}^{+\infty} \phi(z')dz' - \Phi(z) \qquad \text{(normalization)}$$

$$= \int_{-\infty}^{+\infty} \phi(z')dz' - \int_{-\infty}^{z} \phi(z')dz' \qquad \text{(definition)}$$

$$= \int_{-\infty}^{z} \phi(z')dz' + \int_{z}^{+\infty} \phi(z')dz' - \int_{-\infty}^{z} \phi(z')dz' \qquad \text{(integral)}$$

$$= \int_{z}^{+\infty} \phi(z')dz'.$$

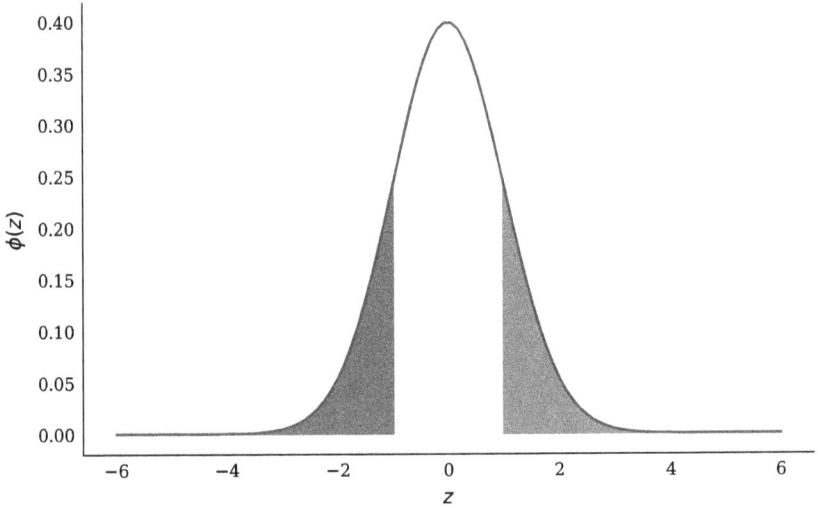

Figure 12.2. Demonstration of the symmetry property of the CDF of the standard Normal. The left and right shaded areas are the same for any z.

Notice that we used the fact that $\phi(z)$ is normalized and a standard property of the integral.

Visually, the expression $\Phi(-z) = 1 - \Phi(z)$ means that the left and the right shaded areas in Figure 12.2 are the same for any z. And this makes a lot of sense since $\phi(z)$ is symmetric. So, in words the property says:

> The probability that Z is smaller than $-z$ is the same as the probability that Z is greater than z.

The formal proof of this property is actually trivial. It goes like this:

$$\Phi(-z) = \int_{-\infty}^{-z} \phi(z')dz'$$

$$= \int_{+\infty}^{z} \phi(\tilde{z})(-1)d\tilde{z},$$

after applying the transformation $\tilde{z} = -z'$. And finally:

$$\int_{+\infty}^{z} \phi(\tilde{z})(-1)d\tilde{z} = -\int_{+\infty}^{z} \phi(\tilde{z})d\tilde{z} = \int_{z}^{+\infty} \phi(\tilde{z})d\tilde{z},$$

which as we saw above is the same as $1 - \Phi(z)$.

Let's demonstrate this in `scipy.stats`:

```
print(f'p(Z <= -1) = {Z.cdf(-1):1.3f}')
print(f'p(Z >= 1) = {1 - Z.cdf(1):1.3f}')
```

You should get:

```
p(Z <= -1) = 0.159
p(Z >= 1) = 0.159
```

What is the probability that Z is between -1 and 1? It is:

$$p(-1 < Z < 1) = 1 - p(Z \leq -1) - p(Z \geq 1) = 1 - 2\Phi(-1).$$

In `scipy.stats`:

```
print(f'p(-1 < Z < 1) = {1 - 2.0 * Z.cdf(-1):1.3f}')
```

You should get:

```
p(-1 < Z < 1) = 0.683
```

12.2 Quantiles of the Standard Normal: Colab

Quantiles are a great way to summarize a random variable with a few numbers. Let's start exploring quantiles with the standard Normal. Take:

$$Z \sim N(0, 1).$$

The most important quantile is the median. The median is the point z that splits the probability of Z into two equal parts. 50% of the probability of Z is to the left of the median and 50% is to the right. So, for the standard Normal, the median is just $z = 0$. The median is also known as the 0.5 quantile. We typically write:

$$z_{0.5} = 0.$$

Of course, `scipy.stats` knows about the median:

```
Z = st.norm()
Z.median()
```

You should get:

```
0.0
```

The definition of a general quantile is this:

> The q quantile of Z is the value z_q such that the probability of Z being less than z_q is q.

Mathematically, you want to find a value z_q such that:

$$\Phi(z_q) = q.$$

To find z_q, you need to solve this equation. In the old days, you would just look it up in a table at the end of your statistics textbook. But now you can use `scipy.stats` to find the quantile:

```
Z.ppf(0.5)
```

The function `ppf` stands for percent point function or inverse CDF. You should get:

```
0.0
```

Another interesting quantile of the standard Normal is $z_{0.025}$. It is the point below which Z lies with probability 2.5%. This is not trivial to find though. You really need to solve the nonlinear equation:

$$\Phi(z_{0.025}) = 0.025.$$

Fortunately, `scipy.stats` can do this for you using the function `Z.ppf()`:

```
import scipy.stats as st
Z = st.norm()
z_025 = Z.ppf(0.025)
print(f'z_025 = {z_025:1.2f}')
```

The output is:

```
z_025 = -1.96
```

Let's verify that this is indeed giving us the 0.025 quantile. If we plug it in the CDF we should get 0.025:

```
print(f'Phi(z_025) = {Z.cdf(z_025):1.3f}')
```

The output is:

```
Phi(z_025) = 0.025
```

Okay, it looks good!

Yet another interesting quantile is $z_{0.975}$. It is the point above which Z lies with probability 97.5%. It is the same as $z_{0.025}$ but with a positive sign.

```
z_975 = Z.ppf(0.975)
print(f'z_975 = {z_975:1.2f}')
```

The output is:

```
z_975 = 1.96
```

Nice! This is just $-z_{0.025}$. We could have guessed it!

12.2.1   *Credible intervals*

The quantiles $z_{0.025}$ and $z_{0.975}$ are particularly important because the probability that Z is between them is 95%. We could write:

Z is between $z_{0.025}$ and $z_{0.975}$ with probability 95%.

This is what we mean by a very nice summary of the uncertainty in Z. This is called the 95% (central) *credible interval* of Z.

Now if you are like me, you would simplify this a bit more by writing:

Z is between -2 and +2 with probability (approximately) 95%.

Who wants to remember that 1.96? Figure 12.3 shows the 95% (central) credible interval of the standard Normal.

Let's end this subsection by finding the 99.9% credible interval of Z. We need the $z_{0.001}$ and $z_{0.999}$ quantiles. We can find them like this:

```
z_001 = Z.ppf(0.001)
print(f'z_001 = {z_001:1.2f}')
z_999 = Z.ppf(0.999)
print(f'z_999 = {z_999:1.2f}')
```

The output is:

```
z_001 = -3.09
z_999 = 3.09
```

So, we can now write:

Z is between -3.09 and 3.09 with probability 99.8%.

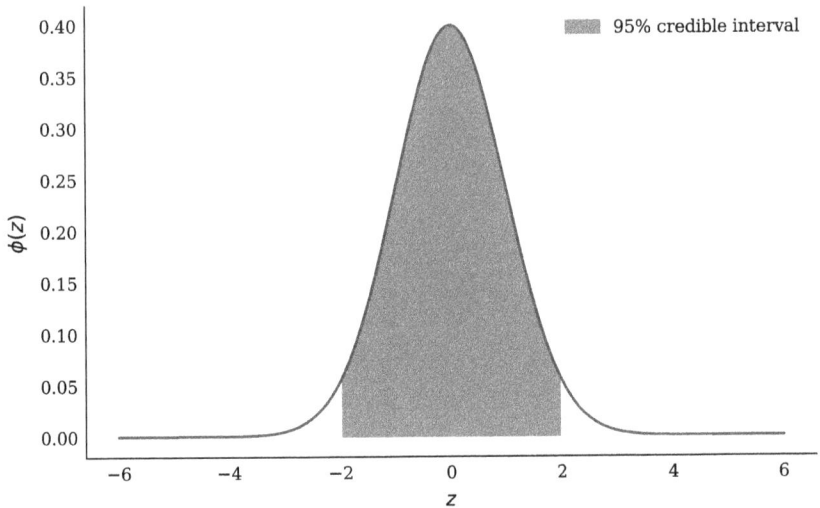

Figure 12.3.   The 95% (central) credible interval of the standard Normal. The shaded area is the probability that Z is between $z_{0.025}$ and $z_{0.975}$. It is 95%.

Or the more practical:

> Z is between -3 and 3 with probability (about) 99.8%.

How can we think about this intuitively? Well, if you sample many many times from Z approximately 2 out of a 1000 samples will be outside of the interval $[-3, 3]$. Let's test this computationally:

```
# Take 1,000,000 samples
zs = Z.rvs(size=1_000_000)
# Count the number of zs that are outside the range
idx = (zs < -3) | (zs > 3)
# How many samples out of 1,000?
zs[idx].size / 1_000_000 * 1_000
```

The output is:

```
2.761
```

12.3   The Normal Distribution: Colab

The Normal (or Gaussian) distribution is a ubiquitous one. It appears over and over again. You must already have seen if you completed all the activities related to the Binomial and the Poisson distributions. There are two explanations as to why the Normal appears so often. First, it is the distribution of maximum uncertainty that matches a known mean and a known variance. This comes from the principle of maximum entropy, a rather advanced concept that we are not going to deal with. Second, it is the distribution that arises when you sum up a lot of independent random variables together. This result is known as the central limit theorem.

We write:

$$X \sim N(\mu, \sigma^2),$$

and we read it as:

X follows a Normal distribution with mean μ and variance σ^2.

Notice that for $\mu = 0$ and $\sigma^2 = 1$, we get the standard Normal. The PDF of X is:

$$p(x) := \frac{1}{\sqrt{2\pi}\sigma} \exp\left\{-\frac{(x-\mu)^2}{2\sigma^2}\right\}.$$

The mean of X is:

$$\mathbf{E}[X] = \mu,$$

and the variance is:

$$\mathbf{V}[X] = \sigma^2.$$

Of course, the standard deviation of X is just σ.

Here is how to define a Normal variable in `scipy.stats`:

```
mu = 5.0
sigma = 2.0
X = st.norm(loc=mu, scale=sigma)
```

We sample as usual. The left side of Figure 12.4 shows the PDF of the Normal distribution with $\mu = 5$ and $\sigma = 2$.

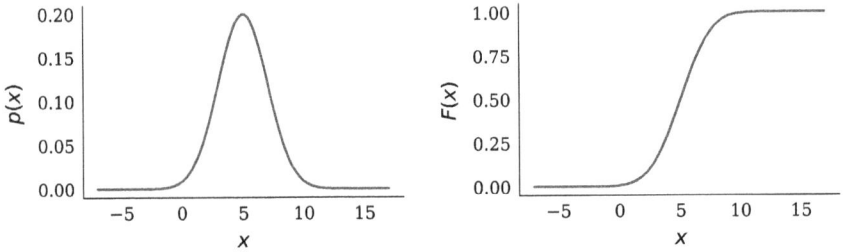

Figure 12.4. The PDF (left) and CDF (right) of the Normal distribution with $\mu = 5$ and $\sigma = 2$.

Notice that the PDF of X is just a scaled version of the standard Normal PDF:

$$p(x) = \frac{1}{\sigma}\phi\left(\frac{x-\mu}{\sigma}\right).$$

This is a very useful observation as it allows us to prove the formula for the mean of X. Indeed, we have:

$$\mathbf{E}[X] = \int_{-\infty}^{+\infty} x p(x) dx$$

$$= \int_{-\infty}^{+\infty} x \frac{1}{\sigma}\phi\left(\frac{x-\mu}{\sigma}\right) dx.$$

Change integration variable to $z = \frac{x-\mu}{\sigma}$, and you get:

$$\mathbf{E}[X] = \int_{-\infty}^{+\infty} (\mu + \sigma z)\frac{1}{\sigma}\phi\left(z\right)\sigma dz$$

$$= \int_{-\infty}^{+\infty} (\mu + \sigma z)\phi\left(z\right) dz$$

$$= \mathbf{E}[\mu + \sigma Z]$$

$$= \mu + \sigma \mathbf{E}[Z]$$

$$= \mu,$$

since $\mathbf{E}[Z] = 0$. Similarly, you can show that $\mathbf{V}[X] = \sigma^2$.

The Normal Distribution, Quantiles, and Credible Intervals 227

There is also a connection between the CDF of X, call it $F(x) = p(X \leq x)$, and the CDF of the standard Normal $\Phi(z)$. It is:

$$F(x) := p(X \leq x) = \Phi\left(\frac{x - \mu}{\sigma}\right).$$

We can formally prove this as follows:

$$F(x) = \int_{-\infty}^{x} p(\tilde{x}) d\tilde{x}$$

$$= \int_{-\infty}^{x} \frac{1}{\sigma} \phi\left(\frac{\tilde{x} - \mu}{\sigma}\right) d\tilde{x}.$$

Now change integration variable to $\tilde{z} = \frac{\tilde{x} - \mu}{\sigma}$. We have that:

$$F(x) = \int_{-\infty}^{\frac{x - \mu}{\sigma}} \frac{1}{\sigma} \exp\left\{-\frac{\tilde{z}^2}{2}\right\} \sigma d\tilde{z}$$

$$= \int_{-\infty}^{\frac{x - \mu}{\sigma}} \phi(\tilde{z}) \tilde{z}$$

$$= \Phi\left(\frac{x - \mu}{\sigma}\right).$$

So, if you plot the CDF of X it looks exactly like the CDF of the standard normal at a different scale. The right side of Figure 12.4 shows the CDF of X.

12.4 Quantiles of the Normal: Colab

Now take a Normal:

$$X \sim N(\mu, \sigma^2),$$

and let $F(x) = p(X \leq x)$ be its CDF. The q-quantile of X is defined through the solution of the non-linear equation:

$$F(x_q) = q.$$

Recall that we have managed to express the Normal CDF $F(x)$ in terms of the standard Normal CDF $\Phi(z)$:

$$F(x) = \Phi\left(\frac{x - \mu}{\sigma}\right).$$

We can use this to express x_q in terms of z_q, i.e., the q-quantile of the standard Normal. We have by substitution in the defining equation of x_q:

$$\Phi\left(\frac{x_q - \mu}{\sigma}\right) = q.$$

By comparing to the defining equation of z_q,

$$\Phi(z_q) = q,$$

we get that:

$$z_q = \frac{x_q - \mu}{\sigma},$$

or, in terms of x_q:

$$x_q = \mu + \sigma z_q.$$

Alright, so if we know the q-quantiles of the standard Normal, we can get the q-quantiles of a Normal. The median is trivial:

$$x_{0.5} = \mu,$$

since $z_{0.5}$. The 0.025-quantile is:

$$x_{0.025} \approx \mu - 1.96\sigma \approx \mu - 2\sigma.$$

And the 0.975-quantile is:

$$x_{0.975} \approx \mu + 1.96\sigma \approx \mu + 2\sigma.$$

So, we can say (95% central credible interval):

X is between $\mu - 2\sigma$ and $\mu + 2\sigma$ with 95% probability.

Similarly, we can find the 0.001-quantile:

$$x_{0.001} \approx \mu - 3.09\sigma \approx= \mu - 3\sigma,$$

and the 0.999-quantile:

$$x_{0.999} \approx \mu + 3.09\sigma \approx= \mu + 3\sigma,$$

which we can use to say (99.8% central credible interval):

X is between $\mu - 3\sigma$ and $\mu + 3\sigma$ with 99.8% probability.

These are good things to remember, but even if you don't, you can still use `scipy.stats` to find quantile and central credible intervals of any Normal:

```
import scipy.stats as st
mu = 5
sigma = 2
X = st.norm(loc=mu, scale=sigma)
x_025 = X.ppf(0.025)
x_975 = X.ppf(0.975)
print(f'X is between {x_025:1.2f} and {x_975:1.2f} with probability
↪    95%')
print('Compare to the interval found through the standard normal')
print(f'** X is between {mu - sigma * 1.96:1.2f} and {mu + sigma *
↪    1.96:1.2f} with probability 95%')
```

The output is:

```
X is between 1.08 and 8.92 with probability 95%
Compare to the interval found through the standard normal
** X is between 1.08 and 8.92 with probability 95%
```

12.5   Fitting Normal Distributions to Data: Colab

Let's say that you have N data points:

$$x_{1:N} = (x_1, \ldots, x_N),$$

and you want to fit a Normal distribution to them. So, you want to find a mean μ and a variance σ^2 so that the data can be thought of as samples from a Normal random variable:

$$X \sim N(\mu, \sigma^2).$$

This is our very first *model training* problem. Our model is that X follows a Normal with unknown mean parameter μ and unknown variance parameter σ^2 and we want to use the data $x_{1:N}$ to find these parameters. This is what we do in data science every day. We train models like this! Sometimes, instead of saying "train a model with" we say that we "fit the model parameters to data." So, "model training" and "parameter fitting" are the same thing.

12.5.1   The method of moments

Okay, how do we fit the parameters of a model? I will show you the simplest possible method. It is called the *method of moments*.[1]

First, what is a *moment* of a random variable X? Well, a moment is this expectation:

$$\mathbf{E}[X^\rho] = \int x^\rho p(x) dx,$$

for any $\rho = 1, 2, \ldots,$. So, the first moment ($\rho = 1$) is just the mean:

$$\mathbf{E}[X] = \mu,$$

and the second moment ($\rho = 2$) is:

$$\mathbf{E}[X^2] = \mathbf{V}[X] + (\mathbf{E}[X])^2 = \sigma^2 + \mu^2,$$

where I used the formula $\mathbf{V}[X] = \mathbf{E}[X^2] - (\mathbf{E}[X])^2$.

The method of moments says that you should pick the parameters of a distribution so that the theoretical moments match the empirical moments. Specifically, the first moment should match the empirical

---

[1]https://en.wikipedia.org/wiki/Method_of_moments_(statistics).

average, i.e.,

$$\mathbf{E}[X] = \frac{1}{N} \sum_{i=1}^{N} x_i,$$

the second moment should match:

$$\mathbf{E}[X^2] = \frac{1}{N} \sum_{i=1}^{N} x_i^2,$$

and so on and so forth.

For the Normal in particular, we have from matching the first moment that we should pick μ to be:

$$\hat{\mu} = \frac{1}{N} \sum_{i=1}^{N} x_i.$$

Note that $\hat{\mu}$ is our estimate of μ. That's what the hat indicates. Similarly, matching the second moment yields:

$$\hat{\sigma}^2 + \hat{\mu}^2 = \frac{1}{N} \sum_{i=1}^{N} x_i^2,$$

which we can solve for $\hat{\sigma}^2$ (our estimate of the variance):

$$\hat{\sigma}^2 = \frac{1}{N} \sum_{i=1}^{N} x_i^2 - \hat{\mu}^2.$$

Okay, let's try it out in an example.

12.5.2   *Example: Fitting a Normal to the selected temperature setpoints*

Let's load our high-performance building dataset:

```
import numpy as np
import scipy.stats as st

# The url of the file we want to download
```

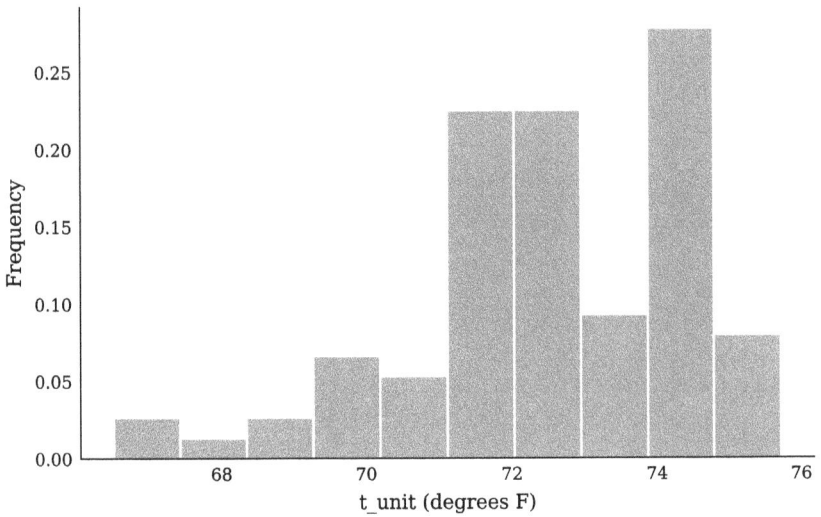

Figure 12.5. The histogram of the selected temperature setpoints for unit a1.

```
# Url is too big to fit in the book
url = '.../temperature_raw.xlsx'
```

```
!curl -O $url
```

```
import pandas as pd
df = pd.read_excel('temperature_raw.xlsx')
df = df.dropna(axis=0)
df.date = pd.to_datetime(df['date'], format='%Y-%m-%d')
```

We are going to pick data from a specific unit, say a1 and we are going to fit a Normal to the selected temperature setpoint t_unit. Select all relevant data:

```
t_unit_a1 = df[df['household'] == 'a1']['t_unit']
```

Figure 12.5 shows the histogram of the selected temperature setpoints. This doesn't look perfectly Normal, but it's okay. We will fit it to a Normal anyway!

Using the method of moments, let's pick the mean:

```
mu = t_unit_a1.mean()
print(f'mu = {mu:1.2f}')
```

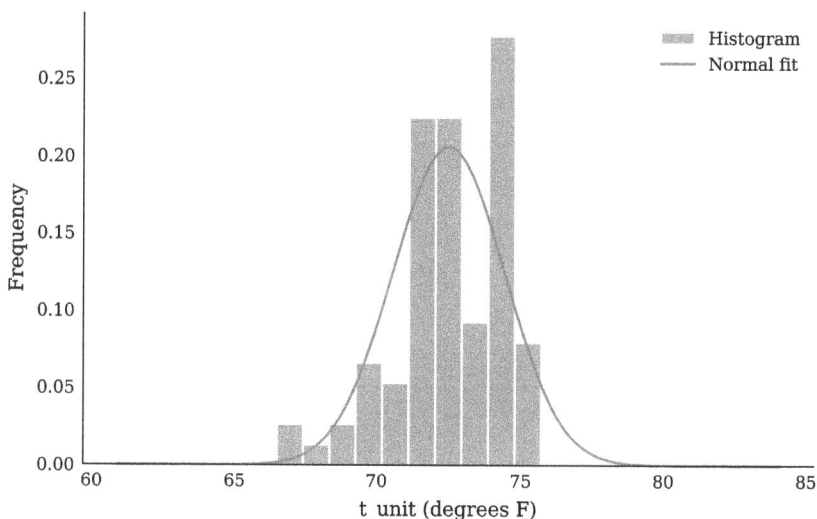

Figure 12.6.   The PDF of $T_{unit,a1}$ and the histogram of the data.

The output is:

```
mu = 72.49
```

and also the variance:

```
sigma2 = np.mean(t_unit_a1 ** 2) - mu ** 2
print(f'sigma2 = {sigma2:1.2f}')
```

The output is:

```
sigma2 = 3.75
```

Now let's make such a random variable:

```
sigma = np.sqrt(sigma2)
T_unit_a1 = st.norm(loc=mu, scale=np.sqrt(sigma))
```

Figure 12.6 shows the PDF of $T_{unit,a1}$ and the histogram of the data.

Okay, not perfect — but not bad either. When you are doing data science you should be able to accept some error in your model. There is no such thing as a perfect model. Here we have managed to summarize the situation with two numbers (mean and variance) which is pretty impressive.

Now that you have fitted a Normal distribution to the data you can ask all sorts of questions. Like:

- What is the 0.025 quantile?
- What is the 0.975 quantile?
- What is the probability that the unit temperature exceeds 72°F?

Here are the answers to these questions (respectively):

```
q025 = T_unit_a1.ppf(0.025)
print(f'0.025-quantile = {q025:1.2f}')
```

Output:

```
0.025-quantile = 68.70
```

```
q975 = T_unit_a1.ppf(0.975)
print(f'0.975-quantile = {q975:1.2f}')
```

Output:

```
0.975-quantile = 76.29
```

```
pgt72 = 1 - T_unit_a1.cdf(72)
print(f'p(T_unit_a1 > 72) = {pgt72:1.2f}')
```

Output:

```
p(T_unit_a1 > 72) = 0.60
```

What else can you do? Well, you can compare the households to each other and try to figure out if there are some that prefer cooler or hotter temperatures. This can be as simple as comparing the means. Let's compare the unit temperatures of a1 and a2 in this way. We have already fitted a Normal to the preferred temperature of a1. Let's do the same thing for a2:

```
t_unit_a2 = df[df['household'] == 'a2']['t_unit']
mu_a2 = t_unit_a2.mean()
sigma2_a2 = np.mean(t_unit_a2 ** 2) - mu_a2 ** 2
sigma_a2 = np.sqrt(sigma2_a2)
T_unit_a2 = st.norm(loc=mu_a2, scale=sigma_a2)
```

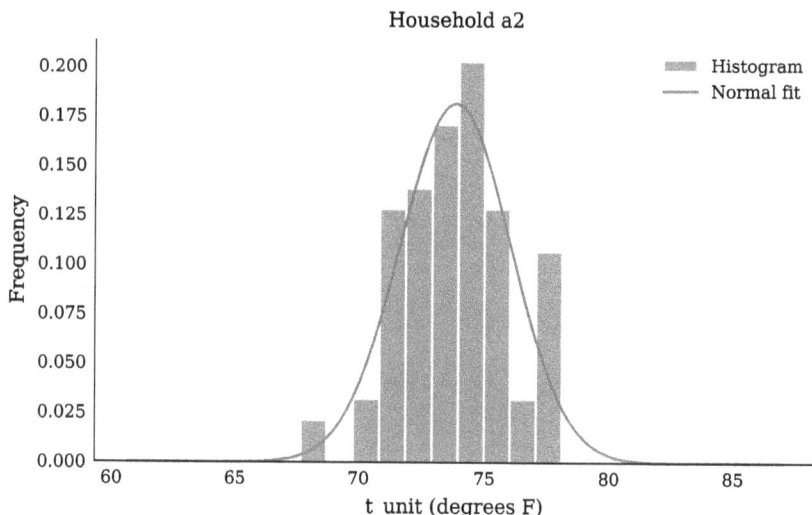

Figure 12.7. The PDF and the histogram for a2.

Figure 12.7 shows the PDF and the histogram for a2.
Alright, let's compare the two means:

```
print(f'mu_a1 = {mu:1.2f}')
print(f'mu_a2 = {mu_a2:1.2f}')
```

Output:

```
mu_a1 = 72.49
mu_a2 = 73.83
```

The estimated mean for the second household is greater. So we can say that the second household prefers warmer temperatures.

12.5.3   *Using Bootstrapping to quantify our uncertainty about the estimated means (Advanced topic)*

Recall our discussion of sampling uncertainty in Section 7.5. If you had a new dataset, you would get a different estimate for the Normal parameters. Is the difference between the means of households a1 and a2 sufficiently big so that you can ignore this uncertainty when you are comparing them? You can use the idea of Bootstrapping to quantify the sampling uncertainty in the estimates of the means. Let's write some code to do exactly that.

```
def find_t_unit_mean_estimate_of_household(household, N, df):
    """Estimates the mean of t_unit by randomly picking N rows
    from the data relevant to the household.

    Arguments:
    household      -      The household name.
    N              -      The number of rows to pick at random.
    df             -      The data frame containing the data.

    Returns: The number of rows with A True divided by N.
    """
    household_df = df[df['household'] == household]
    rows = np.random.randint(0, household_df.shape[0], size=N)
    household_df_exp = household_df.iloc[rows]
    return household_df_exp['t_unit'].mean()
```

Remember that the idea of Bootstrapping is that you replicate the experiment as many times as you want. Here it is 10 times:

```
for i in range(10):
    mu_a1 = find_t_unit_mean_estimate_of_household('a1', 50, df)
    print(f'mu_a1 sample {i} = {mu_a1:1.2f}')
```

Output:

```
mu_a1 sample = 72.19
mu_a1 sample = 72.92
mu_a1 sample = 72.63
mu_a1 sample = 72.24
mu_a1 sample = 73.09
mu_a1 sample = 72.30
mu_a1 sample = 72.68
mu_a1 sample = 72.53
mu_a1 sample = 72.10
mu_a1 sample = 71.76
```

```
for i in range(10):
    mu_a2 = find_t_unit_mean_estimate_of_household('a2', 50, df)
    print(f'mu_a2 sample {i} = {mu_a2:1.2f}')
```

Output:

```
mu_a2 sample = 73.99
mu_a2 sample = 73.62
mu_a2 sample = 73.90
mu_a2 sample = 74.16
mu_a2 sample = 73.59
mu_a2 sample = 73.74
mu_a2 sample = 73.80
mu_a2 sample = 73.67
mu_a2 sample = 73.85
mu_a2 sample = 74.24
```

Now, I'm going to take 1000 sample estimates for each mean and I am going to compare their histograms.

```
mu_a1s = []
mu_a2s = []
for i in range(1000):
    mu_a1s.append(find_t_unit_mean_estimate_of_household('a1', 50,
    ↪  df))
    mu_a2s.append(find_t_unit_mean_estimate_of_household('a2', 50,
    ↪  df))
```

Figure 12.8 shows the histograms of the sample estimates for the means of a1 and a2. And from this, we can clearly see that indeed the mean for a2 is greater than the mean of a1 with very high probability. So, yes a2 prefers warmer temperatures than a1.

Exercises

1. Let Z be the standard Normal. Find the 99.99% central credible interval of Z.
2. Let $X \sim N(5, 2^2)$. Find the 99.9% central credible interval of X.

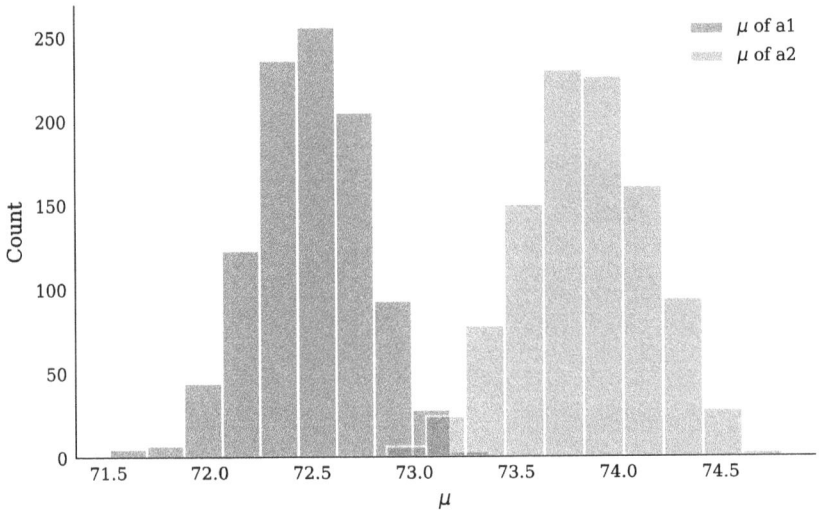

Figure 12.8. The histograms of the sample estimates for the means of a1 and a2.

3. **Comparing the performance of robotic systems:** You are considering purchasing a robotic system for manufacturing masks. There are two possibilities, say A and B. They both produce the same number of masks per day, they cost the same to purchase, and they have the same power and supply costs. However, they are not identical. They have different faulty mask rates. Let X_A and X_B be the number of faulty masks you get from each system, respectively, in a given day. For each of the possibilities below:

   - Use scipy.stats to make two Normal random variables X_A and X_B with the right mean and variance.
   - Plot the PDF of the random variables in the same figure.
   - Find a 95% central credible interval.
   - Indicate which robotic system you would buy and why (three choices A, B, and "I cannot choose").

   (a) **Case 1: $\mathbf{E}[X_A] = 1, \mathbf{V}[X_A] = 0.1$ and $\mathbf{E}[X_B] = 1, \mathbf{V}[X_B] = 0.2$.**

   (b) **Case 2: $\mathbf{E}[X_A] = 1, \mathbf{V}[X_A] = 0.1$ and $\mathbf{E}[X_B] = 2, \mathbf{V}[X_B] = 0.1$.**

   (c) $\mathbf{E}[X_A] = 1, \mathbf{V}[X_A] = 0.3$ and $\mathbf{E}[X_B] = 1.1, \mathbf{V}[X_B] = 0.1$.

4. **Figuring out which household conserves less energy:**
We are going to look at a dataset for which the Normal is not a good fit. In particular, we are going to look at HVAC energy consumption in our high-performance building data. Load the data and process them as we did in the example of this chapter.

(a) Extract the `hvac` column for household `a5`.

(b) Plot the histogram of `hvac` for household `a5`.

(c) Use the method of moments to fit a Normal distribution to the `hvac` data for household `a5`.

(d) In the same figure, show the histogram of `hvac` data for household `a5` (use `density=True` to make sure it is normalized) and the PDF of the Normal you just fitted. Is this a good fit? Why do you think we do not get a good fit?

(e) Now I am asking you to transform the data in a way that will make them look more Normal. Do the histogram of the logarithm of the `hvac` data for household `a5`.

(f) Use the method of moments to fit a Normal to the logarithm of the `hvac` data for household `a5`.

(g) In the same figure, show the histogram of the logarithm of the `hvac` data for household `a5` (use `density=True` to make sure it is normalized) and the PDF of the Normal you just fitted.

(h) Now do exactly the same thing as the previous for household `a3`.

(i) Which household consumes more energy, `a5` or `a3`? How do you know?

5. **The Log-Normal distribution:** In the previous problem, we took the logarithm of `hvac` in order to obtain a better fit to the Normal. It turns out that this is a very common practice whenever you have a positive dataset that is skewed in the way we noticed above. As a matter of fact, there is a distribution called the *Log-Normal distribution*,[2] which is designed to do exactly that. Below, I show you could have fitted a Log-Normal distribution directly on the `hvac` data.

---

[2]https://en.wikipedia.org/wiki/Log-normal_distribution.

```
params = st.lognorm.fit(hvac_a5) # This does something similar
↪    to the method of moments
HVAC_a5_ln = st.lognorm(*params) # This is the random variable
# Here is how you can sample from it:
HVAC_a5_ln.rvs(size=10)
# Here is how you can evaluate its PDF:
HVAC_a5_ln.pdf(200)
```

(a) In the same figure, plot the PDF of the Log-normal along with the histogram of the hvac data for a5.

(b) Is the fit with Log-Normal good?

(c) Recall how in Section 12.2 we used the ppf() function of a random variable to find quantiles. Use this function to find the 0.025-quantile of the Log-Normal we constructed above.

(d) Now find the 95% central credible interval of the Log-Normal variable we constructed above.

(e) Repeat as many code-blocks as you need to fit a Log-Normal to the hvac data for household a3 and then find the 95% central credible interval of the resulting random variable.

(f) By comparing the 95% central credible intervals constructed above, which household a5 or a3 consumes more energy?

Chapter 13

Fitting Models with the Principle of Maximum Likelihood

In this chapter, I explore how we can fit probability models to data using the principle of maximum likelihood. I begin by introducing joint probability mass and density functions, which allow us to model multiple random variables together. I then discuss independent random variables and how they can model the result of multiple experiments. The core of the chapter focuses on maximum likelihood estimation, a powerful method for finding the best parameters for our models. Finally, I show how we can validate our fitted models through predictive checking.

13.1 The Joint PDF

Consider two continuous random variables X and Y. The *joint PDF* of the pair (X, Y) is the function $p(x, y)$ giving the probability density around $X = x$ and $Y = y$. Mathematically, we have

$$p(x, y) \approx \frac{p(x \leq X \leq x + \Delta x, y \leq Y \leq y + \Delta y)}{\Delta x \Delta y},$$

with equality when the box around (x, y) shrinks to a point.

13.1.1   *Properties of the joint PDF*

The joint PDF is nonnegative:

$$p(x, y) \geq 0.$$

If you integrate over all the possible values of all random variables, you should get one:

$$\int p(x, y) dx dy = 1.$$

If you *marginalize* over the values of one of the random variables, you get the PDF of the other. By the word *marginalize*, I mean that you integrate out the variable you are marginalizing over. For example,

$$p(x) = \int p(x, y) dy,$$

and

$$p(y) = \int p(x, y) dx.$$

The integration in the above expressions is not abstract. It is over the possible values of the variable you are marginalizing over. The marginalization property is a direct consequence of the sum rule of probability.

If you take a subset A of $\mathbb{R}^2$, then the probability that (X, Y) is in A is given by

$$p\left((X, Y) \in A\right) = \int_A p(x, y) dx dy.$$

This is a standard double integral.

13.1.2   *Conditioning a random variable on another*

If we had observed that $Y = y$, how would this change the PDF of X? The answer is given via Bayes' rule. The PDF of X conditioned on $Y = y$ is

$$p(x|y) = \frac{p(x, y)}{p(y)}.$$

This is called the conditional PDF of X given $Y = y$. It tells us how the PDF of X changes if we know that $Y = y$.

13.1.3   *Independent random variables*

We say that two random variables X and Y are *independent* if and only if conditioning on one does not tell you anything about the other, i.e.,

$$p(x|y) = p(x).$$

A very important property of independent random variables is that their joint PDF is the product of individual PDFs:

$$p(x, y) = p(x)p(y).$$

You can prove this very easily using the product rule and the definition of independence. Here it is

$$p(x, y) = p(x|y)p(y) = p(x)p(y).$$

It is also very easy to show that if Y does not give any information about X, then the reverse also holds:

$$p(y|x) = p(y).$$

13.1.4   *The joint PDF of more than two random variables*

The concept of joint PDF and independence generalizes to arbitrary number of random variables. For example, you could have $p(x, y, z)$ or $p(x, y, z, w)$, etc. You can do all sorts of things with these objects, e.g., you can marginalize any number of them to get the joint PDF of the rest:

$$p(x, y) = \int p(x, y, z, w) dz dw,$$

or you can condition on one or more of the random variables.

$$p(z, w|x, y) = \frac{p(x, y, z, w)}{p(x, y)}.$$

If the random variables were all independent, then the joint PDF would factorize like this:

$$p(x, y, z, w) = p(x)p(y)p(z)p(w).$$

13.1.5 *Repeated independent experiments*

Imagine that we are performing the same experiment N times. Let's use the random variable X_i to describe the measurement of the ith experiment. Take I to be the background information. This I contains how exactly we prepare the experimental apparatus, the ambient conditions, etc. Now, if we are very careful to prepare everything in exactly the same manner, we can assume that the random variables $X_1, X_2, \ldots, X_N$ are independent conditioned on I. That is, we can assume that the joint PDF factorizes:

$$p(x_1, \ldots, x_N | I) = p(x_1 | I) \ldots p(x_N | I) = \prod_{i=1}^{N} p(x_i | I).$$

Let's introduce the shortcut notation:

$$x_{1:N} = (x_1, \ldots, x_N).$$

We can rewrite the independence property as

$$p(x_{1:N} | I) = \prod_{i=1}^{N} p(x_i | I).$$

Remember this:

> If your data are coming from independent measurements, then their joint PDF factorizes conditioned on the background information.

Why is this important? Well, it is important because it allows you to describe a supercomplicated object like the joint PDF with the product of single-variable PDFs. As we are going to see in the next section, this observation can lead to a very famous and versatile model fitting method: the *maximum likelihood principle*.

Now, in practice, we often do not show the background information I explicitly. We just write $p(x_{1:N})$ instead of $p(x_{1:N} | I)$.

13.2   The Maximum Likelihood Principle

Assume that we have observed some data:

$$x_{1:N} = (x_1, \ldots, x_N),$$

coming from N independent measurements. Let's say that we model its observation with a random variable X_i. Now these random variables will be independent (conditional on the details of the experiment). So, the joint PDF of $X_1, \ldots, X_N$ will factorize.

Furthermore, because the experimental measurements are prepared in the same way, the PDF of each one of the X_i's will be the same function. For example, we can assume that the PDF of X_i is given by

$$p(x_i|\theta) = f(x_i; \theta),$$

where $f(x_i; \theta)$ is some functional form with unknown parameters θ. So, here $f(x; \theta)$ is essentially our model of the experiment and θ are the unknown parameters that we need to determine from the experimental data.[1]

The maximum likelihood principle gives us a way to pick θ using the data $x_{1:N}$. Here is what it is all about. First, write down the joint PDF of the experimental data conditioned on the parameters. It is

$$p(x_{1:N}|\theta) = \prod_{i=1}^{N} f(x_i; \theta).$$

This expression is called the *likelihood of the data* given the model parameters θ. Intuitively, it gives you the probability of observing $x_{1:N}$ if the model parameters were θ.

We are ready for the maximum likelihood principle. The *maximum likelihood principle* tells you that you should pick the parameters that maximize the likelihood. In other words, pick the parameters that make your data most probable under the assumption that the model is correct. So, the training problem becomes now

---

[1] For example, $f(x; \theta)$ could be the PDF of a Normal $N(\mu, \sigma^2)$ distribution and $\theta = (\mu, \sigma^2)$.

an optimization problem:

$$\max_{\theta} p(x_{1:N}|\theta) = \max_{\theta} \prod_{i=1}^{N} f(x_i; \theta).$$

In practice, we maximize the logarithm of the likelihood instead of directly the likelihood, i.e., we solve this optimization problem:

$$\max_{\theta} \log p(x_{1:N}|\theta) = \max_{\theta} \sum_{i=1}^{N} \log f(x_i; \theta);$$

Because the logarithm is an increasing function θ, the two problems are mathematically equivalent. But maximizing the logarithm of the likelihood behaves better numerically.

The estimate of θ you obtain in this way is called the *maximum likelihood estimate* or MLE. Let's now look at a concrete example.

13.3   Fitting the Parameters of a Normal Using the Maximum Likelihood Principle

As before we have N independent measurements. Assume that the measurement X_i follows a Normal distribution with parameters μ and σ^2:

$$p(x_i|\mu, \sigma^2) = f(x_i; \mu, \sigma) = \frac{1}{\sqrt{2\pi\sigma^2}} \exp\left\{-\frac{(x_i - \mu)^2}{2\sigma^2}\right\}.$$

So, in this case:

$$\theta = (\mu, \sigma^2),$$

and

$$f(x; \theta) = \frac{1}{\sqrt{2\pi\sigma^2}} \exp\left\{-\frac{(x - \mu)^2}{2\sigma^2}\right\}.$$

The likelihood of the data is:

$$p(x_{1:N}|\theta) = \prod_{i=1}^{N} f(x_i; \theta)$$

$$= \prod_{i=1}^{N} \frac{1}{\sqrt{2\pi\sigma^2}} \exp\left\{ -\frac{(x_i - \mu)^2}{2\sigma^2} \right\}$$

$$= (2\pi\sigma^2)^{-\frac{N}{2}} \prod_{i=1}^{N} \exp\left\{ -\frac{(x_i - \mu)^2}{2\sigma^2} \right\}$$

$$= (2\pi\sigma^2)^{-\frac{N}{2}} \exp\left\{ -\sum_{i=1}^{N} \frac{(x_i - \mu)^2}{2\sigma^2} \right\}.$$

According to the maximum likelihood principle, we must pick μ and σ^2 that maximizes the logarithm of this expression. Let's find the logarithm first. I am going to call it $J(\mu, \sigma^2)$:

$$J(\mu, \sigma^2) = \log p(x_{1:N}|\theta) = -\frac{N}{2}\log(2\pi) - \frac{N}{2}\log\sigma^2 - \frac{1}{2\sigma^2}\sum_{i=1}^{N}(x_i - \mu)^2.$$

So, now model training has become a calculus problem. You need to maximize the two-variable function $J(\mu, \sigma^2)$ with respect to μ and σ^2. How do you proceed? We could either employ an optimization algorithm or we could do it analytically. Let's do it analytically in this simple case. A necessary condition is that the derivative of J with respect to the parameters is zero. Let's find the derivative of J with respect to μ. It is:

$$\frac{\partial J}{\partial \mu} = \frac{\partial}{\partial \mu}\left[-\frac{N}{2}\log(2\pi) - \frac{N}{2}\log\sigma^2 - \frac{1}{2\sigma^2}\sum_{i=1}^{N}(x_i - \mu)^2 \right]$$

$$= \frac{\partial}{\partial \mu}\left[-\frac{1}{2\sigma^2}\sum_{i=1}^{N}(x_i - \mu)^2 \right]$$

$$= -\frac{1}{2\sigma^2}\sum_{i=1}^{N} \frac{\partial}{\partial \mu}(x_i - \mu)^2$$

$$= -\frac{1}{2\sigma^2} \sum_{i=1}^{N} 2(x_i - \mu) \frac{\partial}{\partial \mu}(x_i - \mu)$$

$$= -\frac{1}{2\sigma^2} \sum_{i=1}^{N} 2(x_i - \mu)(-1)$$

$$= \frac{1}{\sigma^2} \sum_{i=1}^{N} (x_i - \mu)$$

$$= \frac{1}{\sigma^2} \left[\sum_{i=1}^{N} x_i - \sum_{i=1}^{N} \mu \right]$$

$$= \frac{1}{\sigma^2} \left[\sum_{i=1}^{N} x_i - N\mu \right].$$

Alright, that wasn't too bad! Now set this derivative equal to zero and you can solve for μ. You get

$$\hat{\mu} = \frac{1}{N} \sum_{i=1}^{N} x_i.$$

This is nice! It is exactly what we would expect! It also matches what we got using the method of moments. Let's proceed to σ^2. We need to find the derivative of J with respect to it. It is

$$\frac{\partial J}{\partial \sigma^2} = \frac{\partial}{\partial \sigma^2} \left[-\frac{N}{2} \log(2\pi) - \frac{N}{2} \log \sigma^2 - \frac{1}{2\sigma^2} \sum_{i=1}^{N} (x_i - \mu)^2 \right]$$

$$= \frac{\partial}{\partial \sigma^2} \left[-\frac{N}{2} \log(2\pi) - \frac{N}{2} \log \sigma^2 - (\sigma^2)^{-1} \frac{1}{2} \sum_{i=1}^{N} (x_i - \mu)^2 \right]$$

$$= -\frac{N}{2} \frac{\partial}{\partial \sigma^2} \log \sigma^2 - \frac{\sum_{i=1}^{N}(x_i - \mu)^2}{2} \frac{\partial}{\partial \sigma^2} (\sigma^2)^{-1}$$

$$= -\frac{N}{2} \frac{1}{\sigma^2} - \frac{\sum_{i=1}^{N}(x_i - \mu)^2}{2}(-1)(\sigma^2)^{-2}$$

$$= -\frac{N}{2\sigma^2} + \frac{\sum_{i=1}^{N}(x_i - \mu)^2}{2\sigma^4}$$

$$= \frac{-N\sigma^2 + \sum_{i=1}^{N}(x_i - \mu)^2}{2\sigma^4}.$$

Setting this equal to zero and solving for σ^2, yields

$$\hat{\sigma}^2 = \frac{1}{N}\sum_{i=1}^{N}(x_i - \hat{\mu})^2,$$

where I substituted $\hat{\mu}$ for μ. With a little bit of algebra you can show that this is exactly the same as the result obtained with the method of moments, i.e., it can be rewritten as

$$\hat{\sigma}^2 = \frac{1}{N}\sum_{i=1}^{N}x_i^2 - \hat{\mu}^2.$$

Notice that this is the same as the result obtained with the method of moments in Section 12.5.

13.4   Fitting the Bernoulli with Maximum Likelihood

Let's look at another example of maximum likelihood estimation. Assume that we are doing binary experiment with results 1 (success) and 0 (failure). We model the result of the ith experiment using a random variable X_i which follows a Bernoulli distribution with unknown success probability θ, i.e., we assume that:

$$X_i \sim \text{Bernoulli}(\theta).$$

This tells us what is the PMF of each random variable. It is

$$p(X_i = 1|\theta) = \theta,$$

and

$$p(X_i = 0|\theta) = 1 - \theta.$$

There is a common trick that you can use the express this PMF in a single equation. If the experimental result is x_i (either 0 or 1), then we can write the PMF as:

$$p(X_i = x_i|\theta) = \theta^{x_i}(1-\theta)^{1-x_i}.$$

Meditate a bit on this. If $x_i = 1$, then $1 - x_i = 0$ so the second part of the product becomes $(1-\theta)^0 = 1$ and you get θ. Similarly, if $x_i = 0$, the first part of the product becomes $\theta^0 = 1$ and thus you get $1 - \theta$. Nice right? We will see this trick again in this class.

Now do it N times and collect data $x_{1:N}$. Paying attention so that the experiments are independent, we get that the data likelihood is:

$$p(x_{1:N}|\theta) = \prod_{i=1}^{N} p(x_i|\theta).$$

Substituting the expression for the individual PMFs and doing a bit of algebra, we get

$$p(x_{1:N}|\theta) = \prod_{i=1}^{N} p(x_i|\theta)$$

$$= \prod_{i=1}^{N} \theta^{x_i}(1-\theta)^{1-x_i}$$

$$= \theta^{\sum_{i=1}^{N} x_i}(1-\theta)^{\sum_{i=1}^{N}(1-x_i)}$$

$$= \theta^{\sum_{i=1}^{N} x_i}(1-\theta)^{N-\sum_{i=1}^{N} x_i}.$$

This has a nice intuitive meaning. Note that $\sum_{i=1}^{N} x_i$ is the number of successes in N experiments. Similarly, $N - \sum_{i=1}^{N} x_i$ is the number of failures in N experiments.

We want to find the maximum likelihood estimate of the parameter θ. Just like before, we will maximize the logarithm of the likelihood:

$$J(\theta) = \log p(x_{1:N}|\theta) = \sum_{i=1}^{N} x_i \log \theta + \left(N - \sum_{i=1}^{N} x_i\right) \log(1-\theta).$$

We will find the maximum analytically. To this end, take the derivative of J with respect to θ:

$$
\begin{aligned}
\frac{dJ(\theta)}{d\theta} &= \frac{d}{d\theta}\left[\sum_{i=1}^{N} x_i \log\theta + \left(N - \sum_{i=1}^{N} x_i\right)\log(1-\theta)\right] \\
&= \sum_{i=1}^{N} x_i \frac{1}{\theta} + \left(N - \sum_{i=1}^{N} x_i\right)\frac{(-1)}{1-\theta} \\
&= \sum_{i=1}^{N} x_i \frac{1}{\theta} - \left(N - \sum_{i=1}^{N} x_i\right)\frac{1}{1-\theta} \\
&= \frac{\sum_{i=1}^{N} x_i(1-\theta) - (N - \sum_{i=1}^{N} x_i)\theta}{\theta(1-\theta)} \\
&= \frac{\sum_{i=1}^{N} x_i - N\theta}{\theta(1-\theta)}.
\end{aligned}
$$

Setting this to zero and solving for θ yields

$$
\hat{\theta} = \frac{1}{N}\sum_{i=1}^{N} x_i.
$$

Don't you feel good about this result? It's telling you that the observed success frequency is a good estimate for the probability of success![2]

13.5   Predictive Checking: Colab

We saw how we can build models out of data using the maximum likelihood principle. How do we know that these models are good?

---

[2]The MLE becomes better and better as the number of observations increases. As a matter of fact, you can show that if the model is correctly representing reality, the MLE will converge to the true value of the parameters in the limit of infinite samples. In practice, however, you never have infinite samples so MLE will not be perfect. This is fine. You can use Bootstrapping to get estimate of how much uncertainty there is in the MLE estimate. Alternatively, you can use a Bayesian approach to estimate the parameters, albeit this is an advanced topic that goes beyond this course.

Well, this is a very tough question to answer. And, to some extent, it is unanswerable. It is much easier to see when your model is wrong. *Predictive checking* is a way to do exactly that.

Assume that we have built a model using some data. The idea in predictive checking is to use the model to replicate the experiment and compare the results to the real data. If the model is bad, then the replicated data will not match the real data. In that case, you can reject the model. However, you cannot blindly accept a model that performs very well. There is more to it.[3] Predictive checking is good for identifying bugs in your code or coming up with ideas to extend the models in a way that better matches the data than the first idea you had. I will use an example to illustrate this.

13.5.1 *Example of visual predictive checking*

We are investigating a coin toss experiment. Let $x_{1:N}$ be the results of N coin tosses. Each x_i is either 1 (heads) or 0 (tails). We assume that the probability of heads is θ. We model the results of the coin tosses with a Bernoulli distribution:

$$X_i \sim \text{Bernoulli}(\theta).$$

The maximum likelihood estimate of θ is

$$\hat{\theta} = \frac{1}{N} \sum_{i=1}^{N}.$$

The idea is to simply sample $x_{1:N}^{\text{rep}}$ from your fitted model and compare it visually to $x_{1:N}$. Here is how we can replicate the experiment using our fitted model:

- Sample $x^{\text{rep}_{1:N}}$ independently from a Bernoulli with parameter $\hat{\theta}$ (this is one replication of the experiment).
- Repeat as many times as you want. Each time you get a new replication of the experimental data.

---

[3]More specifically, you should consider alternative hypothesis that also replicate the experimental data and then perform (Bayesian) hypothesis testing. This is beyond the scope of this book.

Let's start with a dataset that is 100% compatible with the model. Generate the ground truth data from a fair coin:

```
theta_true = 0.5
X = st.bernoulli(theta_true)
N = 50
xdata = X.rvs(size=N)
xdata
```

A possible output is:

```
array([0, 0, 0, 0, 0, 0, 1, 0, 1, 1, 0, 0, 0, 0, 0, 0, 0, 0, 0, 0,
 ↪   1, 0,
       0, 1, 1, 1, 1, 0, 0, 0, 1, 1, 1, 1, 0, 0, 1, 1, 0, 1, 0, 1,
       ↪   1, 0,
       1, 0, 1, 1, 0, 0])
```

Now find the MLE for θ:

```
hat_theta = xdata.mean()
print(f'hat theta = {hat_theta:1.2f}')
```

Output:

```
hat theta = 0.40
```

Now that we have fitted the model, let's draw many replicated data $x_{1:N}^{\text{rep}}$ from it and compare them visually to the original dataset.

```
# Number of replications
N_rep = 9
# Array to store the replicated experiments
# each row is a replication of the experimental data
x_rep = np.ndarray((N_rep, N))
for i in range(N_rep):
    x_rep[i, :] = st.bernoulli(hat_theta).rvs(size=N)
```

I am not showing the output here because it is a bit long. We can visualize the experimental and the replicated data as images. See the top panel of Figure 13.1. In this image, the first row of pixels is are the observed data. The other rows are the replicated data. The replicated data is visually indistinguishable from the observed data. This is what we would expect if the model is correct.

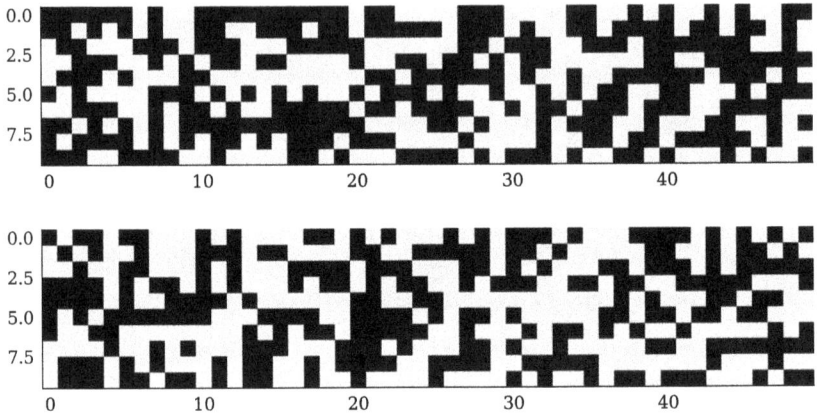

Figure 13.1. Top: visualization of the actual experimental data (first row) followed by 9 replications of the experiment using the fitted model. Bottom: visualization of the synthetic data and 9 replications of the experiment.

Let's now repeat this exercise with a dataset that is fabricated and does not match the model. As a matter of fact, I just picked the dataset by hand trying to get an equal number of 0's and 1's. Can predictive checking reveal that this dataset was not generated by a Bernoulli experiment? Let's see. Here is the data I came up with:

```
xdata_2 = np.array([0, 1, 0, 0, 1, 0, 0, 1, 1, 1, 0, 1, 0, 1, 1, 1,
 ↪  1, 0, 0, 1, 0, 1, 0, 1, 1, 1, 0, 1, 0, 1, 0, 0, 0, 1, 0, 1, 1,
 ↪  1, 0, 0, 0, 0, 1, 0, 1, 0, 1, 0, 1, 0])
```

This is the maximum likelihood estimator for this new coin:

```
hat_theta2 = xdata_2.mean()
print(f'hat theta 2 = {hat_theta2:1.2f}')
```

Output:

```
hat theta 2 = 0.50
```

Let's replicate the experiment 9 times:

```
x_rep_2 = np.ndarray((N_rep, N))
for i in range(N_rep):
    x_rep_2[i, :] = st.bernoulli(hat_theta2).rvs(size=N)
x_rep_2
```

The bottom panel of Figure 13.1 shows the fabricated data and 9 replications of the experiment as an image. Does the first row look similar to the rest of the rows? Well, if you pay close attention, you will notice that the fabricated dataset has more transitions from heads to tails than the replicated data. In other words, the replicated data seem to have longer consecutive series of either heads or tails. How can we see this more clearly? We can use *test quantities* to quantify this.

13.5.2   *Example of test quantities*

We use *test quantities* to characterize the discrepancy between the model and the data. Test quantities help us zoom into the characteristics of the data that are of particular interest to us.

Mathematically, a test quantity is a scalar function of the data and the model parameters $T(x_{1:n}, \theta)$. There are some general recipes for creating test quantities for certain types of models. However, in general, you are better off using common sense in selecting them. Think: what are the important characteristics of the data that your model should be capturing? We will be seeing specific examples below.

Now, assume that you have selected a test quantity $T(x_{1:n}, \theta)$. What do you do with it? Well, the easiest thing to do is a visual comparison of the histogram of the test quantity over replicated data, and compare it to the observed value $T(x_{1:n}, \theta)$. In these plots, you are trying to see how likely or unlikely is the observed test quantity under the assumption that your model is correct. You need an example to understand this.

Let's continue with the coin-toss example of the previous subsection. What is a good test quantity? An obvious one is the number of heads. This is only a function of the data. It is: $T_h(x_{1:n}) = \sum_{i=1}^{n} x_i$.

Let's implement this as a Python function of the data:

```python
def T_h(x):
    """This is an implementation of a
    """
    return x.sum()
```

We are going to try two datasets: the one that was generated from the correct model, and the one I picked by hand. We will see the results that we get from both. For the first dataset (the one generated by the model) we get:

```
# The observed test quantity
T_h_obs = T_h(xdata)
print(f'The observed test quantity is {T_h_obs:d}')
# Draw replicated data
N_rep = 1000
x_rep = np.ndarray((N_rep, N))
for i in range(N_rep):
    x_rep[i, :] = st.bernoulli(hat_theta).rvs(size=N)
# Evaluate the test quantity
T_h_rep = np.ndarray(x_rep.shape[0])
for i in range(x_rep.shape[0]):
    T_h_rep[i] = T_h(x_rep[i, :])
```

Output:

```
The observed test quantity is 20
```

Let's plot the histogram of the replicated test quantity and compare it to the observed test quantity.

```
fig, ax = plt.subplots()
tmp = ax.hist(T_h_rep, density=True, alpha=0.25, label='Replicated
    test quantity')[0]
ax.plot(T_h_obs * np.ones((50,)), np.linspace(0, tmp.max(), 50),
    'k', label='Observed test quantity')
plt.legend(loc='best')
plt.title('Real data, test quantity: number of heads')
```

The result is shown in the left panel of Figure 13.2.

Now, let's look also at the other dataset (the one I picked by hand). The replication code is similar to the one we used for the real data. The only difference is that you should use the fabricated data and the corresponding MLE. The result is shown in the right panel of Figure 13.2.

The two plots in Figure 13.2 look about the same. This just means that I was able to replicate this particular test quantity when I picked values by hand. It is not possible to see the difference with this test statistic.

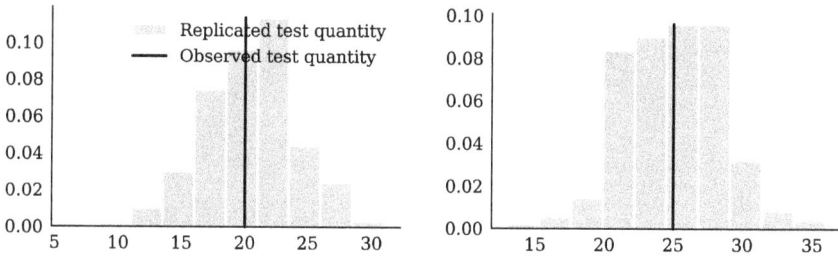

Figure 13.2. Test quantity: number of heads. Left: histogram of the replicated test quantity and the observed test quantity for the real data. Right: histogram of the replicated test quantity and the observed test quantity for the hand-picked data.

Can we find a better test statistic? Something that can discriminate between the two data generating processes? Remember our observation when we plotted the replicated data versus the true data for the second case. We observed that the fabricated data included more transitions from heads to tails. I was trying too hard make them look random. Let's build a statistic that captures that. Here is the one I came up with:

$$T_s(x) = \# \text{ number of switches from 0 and 1 in the sequence } x.$$

This is not easy to write in an analytical form, but we can code it:

```
def T_s(x):
    s = 0
    for i in range(1, x.shape[0]):
        if x[i] != x[i-1]:
            s += 1
    return s
```

The left panel of Figure 13.3 shows the histogram of the replicated test quantity and the observed test quantity for the real data. It looks as expected. The observed test quantity is highly probable under the assumptions of the model. Now, look at the right panel of Figure 13.3. It shows the histogram of the replicated test quantity and the observed test quantity for the hand-picked data. Notice that the observed test quantity is highly unlikely under the assumptions of the model. You see how predictive checking can help you reveal that there is something fishy with your data.

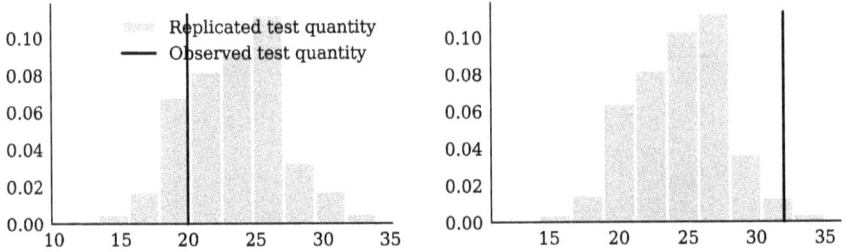

Figure 13.3. Test quantity: number of transitions. Left: histogram of the replicated test quantity and the observed test quantity for the real data. Right: histogram of the replicated test quantity and the observed test quantity for the hand-picked data.

What we did here has connections to the concept of p-values, albeit it is not the same thing. Keep in mind that according to the American Statistical Association, p-values can be easily misused (Wasserstein and Lazar, 2016). The problem is that scientists and engineers tend to use p-values to automatically reject/accept hypotheses. Things are a bit more nuanced than that. To do a good job teaching p-values I should devote at least one chapter on them, and this would be at the expense of another topic. Furthermore, you can do excellent data science without ever using p-values! These are the reasons why I have chosen not to talk about them. If you want to learn about p-values, I suggest you take a core statistics course.

Exercises

1. **Maximum likelihood estimate of an Exponential random variable**
   You are managing toll station and your job is to come up with a probabilistic model for the time that passes between car arrivals. You have at your disposal data:

$$T_{1:N} = (t_1, t_2, \ldots, t_N).$$

Here t_1 is the time that passed until the first event, t_2 is the time that passed from the first event to the second, and so on. You decide to model these random time intervals using an Exponential distribution with unknown rate parameter λ, i.e.,

$$T_i | \lambda \sim \text{Exponential}(\lambda).$$

In terms of the PDF:

$$p(t_i | \lambda) = \lambda e^{-\lambda t_i}.$$

You have decided to use maximum likelihood to fit the parameter λ. Answer the following questions:

(a) Find the mathematical form of the likelihood of the data $p(t_{1:N} | \lambda)$.
(b) Find the mathematical form of the log-likelihood $J(\lambda) = \log p(t_{1:N} | \lambda)$.
(c) Take the derivative $\frac{dJ(\lambda)}{d\lambda}$ of $J(\lambda)$ with respect to λ.
(d) Find the MLE $\hat{\lambda}$ by solving the equation $\frac{dJ(\lambda)}{d\lambda} = 0$ for λ.

2. **Failure of a mechanical component**
Assume that you are designing a gear for a mechanical system. Under normal operating conditions, the gear is expected to fail at a random time. Let T be a random variable capturing the time the gear fails. What should the probability density of T look like? Well, when the gear is brand new, the probability density should be close to zero because a new gear does not fail under normal operating conditions. As time goes by, the probability density should increase because various things start happening to the material, e.g., crack formation, fatigue, etc. Finally, the probability density must again start going to zero as time further increases because nothing lasts forever... A probability distribution that is commonly used to model this situation is the Weibull distribution. We are going to fit some fail time data to a Weibull distribution and then you will have to answer a few questions about failing times. Here is the data:

```
import numpy as np
# Time to fail in years under normal operating conditions
# Each row is a different gear
time_to_fail_data = np.array([
```

```
    10.5,
    7.5,
    8.1,
    8.4,
    11.2,
    9.3,
    8.9,
    12.4
])
```

Here is a Weibull distribution fitted to the data using maximum likelihood:

```
import scipy.stats as st
fitted_params = st.exponweib.fit(time_to_fail_data, loc=0)
T = st.exponweib(*fitted_params)
```

(a) Plot the PDF of the fitted Weibull distribution.
(b) Find the mean fail time and its variance. *Hint*: Do not integrate anything by hand. Just use the functionality of scipy.stats.
(c) Plot the CDF of the fitted Weibull distribution.
(d) Plot the probability that gear survives for more than t as a function of t. That is, plot the function:

$$S(t) = p(T > t).$$

Hint: First express this function in terms of the cumulative distribution function of T.

(e) Find the probability that the gear lasts anywhere between 8 and 10 years.
(f) If you were to sell the gear, how many years "warranty" would you offer and why?

Chapter 14

Covariance, Correlation, and Linear Regression with One Variable

In this chapter, I introduce the concepts of covariance and correlation, and I explain why correlation does not imply causation. I also show how the correlation coefficient can miss non-linear relationships between two quantities. I present the regression problem. I then introduce the linear model, develop the least squares approach for fitting linear models, derive its mathematical solution, and discuss how to validate a regression model using diagnostic tools.

14.1 Covariance between Two Random Variables: Colab

The concept of covariance summarizes with a single number how two random variables X and Y vary together. There are three possibilities of how X and Y can vary together:

- if X is increased, then Y will likely increase,
- if Y is decreased, then X will likely decrease, and
- X and Y are not linked.

Before defining the concept of covariance mathematically, let's load the smart buildings dataset which will help us demonstrate the

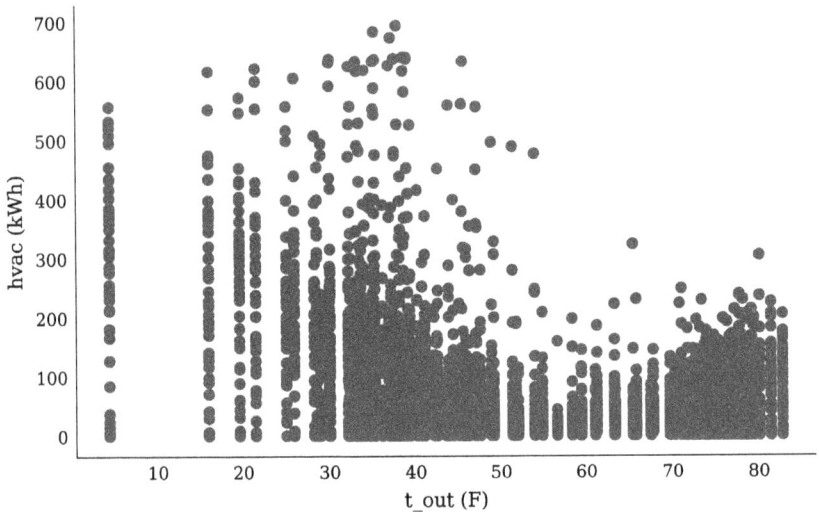

Figure 14.1. Scatter plot of `hvac` (consumed HVAC energy in kWh) and `t_out` (external temperature in degrees F).

concept. The relevant file is temperature_raw.xlsx and you can load it using the following code:

```
import pandas as pd
df = pd.read_excel('temperature_raw.xlsx')
df = df.dropna(axis=0)
```

Figure 14.1 shows the scatter plot of `hvac` (consumed HVAC energy in kWh) and `t_out` (external temperature in degrees F). We see three clear regions here, heating, cooling, and off. Let me separate the data in different dataframes corresponding to these three regions.

```
df_heating = df[df['t_out'] < 60]
df_cooling = df[df['t_out'] > 70]
df_off = df[(df['t_out'] >= 60) & (df['t_out'] <= 70)]
```

Figure 14.2 shows the scatter plots of `hvac` and `t_out` for the three regions. Each region is shown in a different color. The covariance and the correlation will allow us to characterize the relationship between $X = $ t_out and $Y = $ hvac in each one of these regions with a single

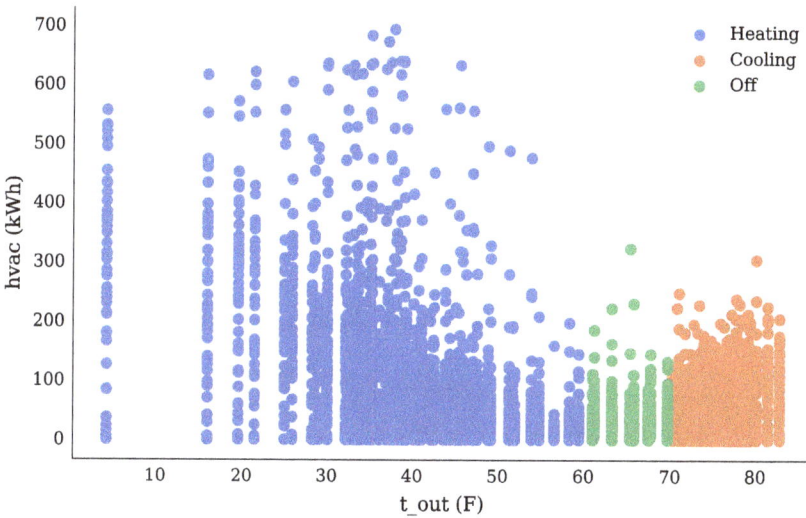

Figure 14.2. Scatter plots of `hvac` and `t_out` for the three regions. Each region is shown in a different color.

number. Depending on the sign of this number (positive, negative, or zero), we can say how the relationship between X and Y goes. In these three regions, we find:

- Heating region (t_out < 60 F): In this regime, increasing $X =$ t_out decreases energy consumption $Y =$ hvac because you use less heating. In the mathematical jargon, we say that X and Y are *negatively correlated*.
- Cooling region (t_out > 70 F): In this regime, increase $X =$ t_out increases energy consumption $Y =$ hvac because you use more cooling. In the mathematical jargon, we say that X and Y are *positively correlated*.
- Off region (60 F $\leq$ t_out $\leq$ 70 F): In this regime, the $X =$ t_out does not affect energy consumption because the HVAC is most likely off. In the mathematical jargon, we say that X and Y are *uncorrelated*.

I am going to do two things next. I will first give you the mathematical definition of covariance and correlation and second I will show you how to estimate them from the data we have. Let's go.

14.1.1 *Mathematical definition of covariance*

Let $p(x, y)$ be the joint PDF of the random variables X and Y. We may or we may not know this, but it certainly exists. Now, let

$$\mu_X = \mathbf{E}[X],$$

be the mean of X and

$$\mu_Y = \mathbf{E}[Y],$$

be the mean of Y. The covariance of X and Y is defined to be:

$$\mathbf{C}[X, Y] := \mathbf{E}\left[(X - \mu_X)(Y - \mu_Y)\right].$$

So, it is the expectation of the product $(X - \mu_X)(Y - \mu_Y)$. Why is this a good definition of how X and Y vary together? To develop your intuition about it, let's look at what the covariance turns out to be in three specific cases:

14.1.1.1 *Case 1: If X and Y are independent, then the covariance is zero*

Let's assume that X and Y are independent. Then, their joint PDF would factorize:

$$p(x, y) = p(x)p(y).$$

This can be exploited to show that $\mathbf{C}[X, Y]$ would be exactly zero. Here it is:

$$\mathbf{C}[X, Y] = \mathbf{E}\left[(X - \mu_X)(Y - \mu_Y)\right]$$

$$= \int (x - \mu_X)(y - \mu_Y)p(x, y)dxdy$$

$$= \int (x - \mu_X)(y - \mu_Y)p(x)p(y)dxdy \text{ (independence)}$$

$$= \int (x - \mu_X)p(x)dx \int (y - \mu_Y)p(y)dy \text{ (calculus)}$$

$$= \mathbf{E}[X - \mu_X] \cdot \mathbf{E}[Y - \mu_Y] \text{ (definition of expectation)}$$

$$= 0 \cdot 0$$

$$= 0$$

14.1.1.2   *Case 2: If $Y = aX + b$ for some positive constant a, then the covariance is positive*

Let's assume that there is a very simple relationship between X and Y:

$$Y = aX + b,$$

for some a positive, and an arbitrary b. This is the simplest way in which an increase in X would yield and increase in Y. Let's see what covariance we get in this case. Notice that the mean of Y is now:

$$\mu_Y = \mathbf{E}[Y] = \mathbf{E}[aX + b] = a\mathbf{E}[X] + b = a\mu_X + b.$$

So, the covariance is:

$$
\begin{aligned}
\mathbf{C}[X, Y] &= \mathbf{E}\left[(X - \mu_X)(Y - \mu_Y)\right] \\
&= \mathbf{E}\left[(X - \mu_X)(aX + b - a\mu_X - b)\right] \\
&= \mathbf{E}\left[(X - \mu_X)a(X - \mu_X)\right] \\
&= \mathbf{E}\left[a(X - \mu_X)^2\right] \\
&= a\mathbf{E}\left[(X - \mu_X)^2\right] \\
&= a\mathbf{V}[X],
\end{aligned}
$$

which is, of course, positive because both a and the variance of X are positive.

14.1.1.3   *Case 3: If $Y = -aX + b$ for some positive constant a, then the covariance is negative*

Let's assume that there is a very simple relationship between X and Y:

$$Y = -aX + b,$$

for some a positive, and an arbitrary b. This is the simplest way in which an increase in X would yield and decrease in Y. In exactly the same way as before, we can show that:

$$\mathbf{C}[X, Y] = -a\mathbf{V}[X],$$

which is a negative number.

14.1.2   *Empirical estimation of the covariance*

Alright, so the covariance does have the intuitive meaning that we want. But, how can we find it if we do not know the joint PDF $p(x, y)$. We will show how you can estimate it from samples of X and Y? So, let's say that we have N measurements of X and Y, say (x_i, y_i) for $i = 1, \ldots, N$. We need the means, which we already know how to estimate:

$$\hat{\mu}_X = \frac{1}{N} \sum_{i=1}^{N} x_i,$$

and

$$\hat{\mu}_Y = \frac{1}{N} \sum_{i=1}^{N} y_i,$$

Okay, we need to estimate one more expectation. Let's do it:

$$
\begin{aligned}
\mathbf{C}[X, Y] &= \mathbf{E}\left[(X - \mu_X)(Y - \mu_Y)\right] \\
&\approx \mathbf{E}\left[(X - \hat{\mu}_X)(Y - \hat{\mu}_Y)\right] \quad \text{(replacing means with estimates)} \\
&\approx \frac{1}{N} \sum_{i=1}^{N} (x_i - \hat{\mu}_X)(y_i - \hat{\mu}_Y)
\end{aligned}
$$

(sampling estimate of expectation).

So, here is our estimate of the covariance[1]:

$$\hat{\sigma}_{X,Y} = \frac{1}{N} \sum_{i=1}^{N} (x_i - \hat{\mu}_X)(y_i - \hat{\mu}_Y).$$

---

[1]The standard estimate of covariance differs a bit from what I have above. It is usually estimated by:

$$\tilde{\sigma}_{X,Y} = \frac{1}{N-1} \sum_{i=1}^{N} (x_i - \hat{\mu}_X)(y_i - \hat{\mu}_Y).$$

This is a so-called *unbiased estimator*. However, if N is big enough the difference is negligible and we don't have to worry about it.

14.1.3 *Example: Covariance between t_out and hvac during heating, cooling, and off*

Let's now calculate the estimate we developed for the covariance in the smart buildings dataset. In particular, we are going to estimate the covariance between $X = $ t_out and $Y = $ hvac for the three regions considered.

Fortunately, we do not have to calculate it by hand. We can use built-in functionality of np.cov. Here is how. Let's do first cooling.

```
import numpy as np
C = np.cov(df_cooling['t_out'], df_cooling['hvac'])
print(C)
```

We get:

```
[[    9.74568849    36.85230047]
 [   36.85230047  2299.42112855]]
```

Let me explain to you what np.cov returns in our case. First, you notice we have returns a 2×2 matrix C. The diagonal of that matrix include the variances of the two datasets. So, here C[0,0] is the variance of df_cooling['t_out']. Check this out:

```
print(f"Variance of df_cooling['t_out'] =
↪    {df_cooling['t_out'].var():1.2f}")
print(f"Compare to C[0, 0] = {C[0, 0]:1.2f}")
```

Output:

```
Variance of df_cooling['t_out'] = 9.75
Compare to C[0, 0] = 9.75
```

Similarly, C[1,1] is the variance of df_cooling['hvac']:

```
print(f"Variance of df_cooling['hvac'] =
↪    {df_cooling['hvac'].var():1.2f}")
print(f"Compare to C[1, 1] = {C[1, 1]:1.2f}")
```

Output:

```
Variance of df_cooling['hvac'] = 2299.42
Compare to C[1, 1] = 2299.42
```

Now `C[0, 1]` is the covariance between the first input $(0 = \texttt{t\_out})$ and the second input $(1 = \texttt{hvac})$. Here it is:

```
print(f"C['t_out', 'hvac'|cooling] = {C[0, 1]:1.2f}")
```

Output:

```
C['t_out', 'hvac'|cooling] = 36.85
```

This is positive for cooling as we expected. Increasing `t_out` results in increasing `hvac`.

But what is `C[1, 0]`? Well, this is the covariance between the second input $(1 = \texttt{hvac})$ and the first input $(0 = \texttt{t\_out})$:

```
print(f"C['hvac', 't_out'|cooling] = {C[1, 0]:1.2f}")
```

Output:

```
C['hvac', 't_out'|cooling] = 36.85
```

This is exactly the same as `C[0, 1]`. Of course, this is not an accident. The covariance between two random variables is a symmetric operator, i.e.,

$$\mathbf{C}[X, Y] = \mathbf{C}[Y, X].$$

The proof of this statement is trivial. Just look at the definition of covariance.

Alright, let's now look at the heating covariance:

```
C = np.cov(df_heating['t_out'], df_heating['hvac'])
print(C)
```

We get:

```
[[  105.27235776  -525.43752907]
 [ -525.43752907 12967.67912181]]
```

```
print(f"C['hvac', 't_out'|heating] = {C[1, 0]:1.2f}")
```

Output:

```
C['hvac', 't_out'|heating] = -525.44
```

It is a nice negative number. Again, this is compatible with our intuition. Negative means that if t_out is increased, hvac decreases. That's exactly what should be happening during heating.

Let's do the off regime:

```
C = np.cov(df_off['t_out'], df_off['hvac'])
print(C)
```

We get:

```
[[   8.08975479   -2.65681839]
 [  -2.65681839 1306.35076875]]
```

```
print(f"C['hvac', 't_out'|off] = {C[1, 0]:1.2f}")
```

Output:

```
C['hvac', 't_out'|off] = -2.66
```

This is smaller in absolute value than any of the other covariance. But it is still negative. Is this -2.66 small? Or is it big? How do we know? Well, that is what the correlation coefficient is going to help us decide.

14.2   Correlation between Two Random Variables: Colab

The covariance between two random variables X and Y, $\mathbf{C}[X,Y]$, is not an absolute measure. As a matter of fact, the covariance has units of X times units of Y. So, if you change the units of X and Y, the covariance will change. Changing units of X is like defining a new random variable:

$$X' = \lambda X.$$

The covariance between X' and Y would be:

$$\begin{aligned}
\mathbf{C}[X',Y] &= \mathbf{E}[(X' - \mu_{X'})(Y - \mu_Y)] \\
&= \mathbf{E}[(\lambda X - \lambda \mu_X)(Y - \mu_Y)] \\
&= \lambda \mathbf{E}[(X - \mu_X)(Y - \mu_Y)] \\
&= \lambda \mathbf{C}[X,Y].
\end{aligned}$$

As an example, imagine that X was measured in meters and you wanted to change it units to centimeters. Then $\lambda = 100$ and $X' = 100X$. The covariance between the new variable X' and Y would be 100 times bigger! As I said, the covariance is not an absolute measure.

How can we fix this? Well, we fix it using the concept of correlation. It is going to be a number that is independent of the units of X and Y. It will be between -1 and 1.

14.2.1 *Mathematical definition of the correlation coefficient*

The *correlation coefficient* between two random variables X and Y is defined to be the covariance between the two random variables divided by the product of their standard deviations, i.e.,

$$\rho(X,Y) = \frac{\mathbf{C}[X,Y]}{\sigma_X \sigma_Y}.$$

where

$$\sigma_X = \sqrt{\mathbf{V}[X]},$$

and

$$\sigma_Y = \sqrt{\mathbf{V}[Y]}.$$

Alright, let's see why this is a good measure by looking closely at some of its properties.

14.2.1.1 *Property 1: The correlation coefficient remains unchanged when you change the units of the random variables*

Okay, this sounds good. Take the example we gave above:

$$X' = \lambda X.$$

We have already shown that:

$$\mathbf{C}[X',Y] = \mathbf{C}[\lambda X, Y] = \lambda \mathbf{C}[X,Y].$$

Notice that:

$$\sigma_{X'} = \sqrt{\mathbf{V}[X']} = \sqrt{\mathbf{V}[\lambda X]} = \sqrt{\lambda^2 \mathbf{V}[X]} = \lambda\sqrt{\mathbf{V}[X]} = \lambda\sigma_X.$$

So, we have:

$$
\begin{aligned}
\rho(X', Y) &= \frac{\mathbf{C}[X', Y]}{\sigma_{X'}\sigma_Y} \\
&= \frac{\lambda\mathbf{C}[X, Y]}{\lambda\sigma_X\sigma_Y} \\
&= \frac{\mathbf{C}[X, Y]}{\sigma_X\sigma_Y} \\
&= \rho(X, Y).
\end{aligned}
$$

That's it. The correlation coefficient has absolute meaning.

14.2.1.2 *Property 2: Two independent random variables have zero correlation*

This property follows directly from the fact that the covariance between two independent random variables is zero.

14.2.1.3 *Property 3: The maximum possible correlation between two random variables is one*

This is a nice property! Why does it hold? Take a random variable X. What is the random variable Y that is the most correlated with X? Well, can you think of something more correlated than $Y = X$? I don't think so... Let's see what correlation coefficient we get when we plug in $Y = X$:

$$
\begin{aligned}
\rho(X, X) &= \frac{\mathbf{C}[X, X]}{\sigma_X\sigma_X} \\
&= \frac{\mathbf{E}[(X - \mu_X)(X - \mu_X)]}{\sigma_X^2} \\
&= \frac{\mathbf{E}[(X - \mu_X)^2]}{\sigma_X^2} \\
&= \frac{\mathbf{V}[X]}{\sigma_X^2}
\end{aligned}
$$

$$= \frac{\sigma_X^2}{\sigma_X^2}$$

$$= 1.$$

Great! By the way, notice that we also showed that the covariance of a random variable with itself is the variance.

14.2.1.4   Property 4: The minimum possible correlation
             between two random variables is minus one

This property is proved in a similar manner the previous one. What is the random variable Y that is most negatively correlated with X? It is $Y = -X$. If you plug this in the correlation formula, you will get:

$$\rho(X, -X) = -1.$$

14.2.2   Summarizing the properties of the correlation coefficient

So, the correlation is a much better measure than the covariance when you want to assess how two random variables vary together. You have the following possibilities:

- If the correlation is zero, then the two random variables are uncorrelated. This doesn't mean that they are independent though. It just means that they *may be* independent. We will elaborate on this later.
- The closer the correlation coefficient is to plus one, the more positively correlated the random variables are.
- The closer the correlation coefficient is to minus one, the more negatively correlated the random variables are.

14.2.3   Empirical estimation of the correlation coefficient

In this section, we are going to show how you can estimate the correlation coefficient from samples of the two random variables. I have already told you how you can estimate the covariance. We also know how to estimate the standard deviations with averages:

$$\hat{\sigma}_X = \sqrt{\frac{1}{N}(x_i - \hat{\mu}_X)^2},$$

and

$$\hat{\sigma}_Y = \sqrt{\frac{1}{N}(y_i - \hat{\mu}_Y)^2}.$$

So, our estimate for the correlation coefficient is:

$$\hat{\rho}_{X,Y} = \frac{\hat{\sigma}_{X,Y}}{\hat{\sigma}_X \hat{\sigma}_Y}.$$

14.2.4 Example: Correlation between t_out and hvac during heating, cooling, and off

Let's now calculate the estimate we developed for the correlation coefficient in the smart buildings dataset. In particular, we are going to estimate the correlation coefficient between X = t_out and Y = hvac for the three regions considered.

Again, we do not have to calculate anything by hand. We can use built-in functionality of np.corrcoef. Here is how. We start with cooling.

```
rho = np.corrcoef(df_cooling['t_out'], df_cooling['hvac'])
rho
```

Output:

```
[[1.         0.24617791]
 [0.24617791 1.        ]]
```

Notice that this is a matrix as well. It has the same format as the matrix returned by np.cov:

- rho[0, 0] is the correlation coefficient between the first input (0 = t_out) and the first input (0 = t_out). So it has to be one always.
- rho[1, 1] is the correlation coefficient between the second input (1 = hvac) and the second input (1 = hvac). Again, it is always one.
- rho[0, 1] is the correlation coefficient between the first input (0 = t_out) and the second input (1 = hvac).

274 Introduction to Data Science for Engineering Students

- `rho[1, 0]` is the correlation coefficient between the second input (1 = `hvac`) and the first input (0 = `t_out`). This is because the correlation coefficient is also symmetric.

Okay. Here is what we were after:

```
print(f"rho['t_out', 'hvac'|cooling] = {rho[0, 1]:1.2f}")
```

Output:

```
rho['t_out', 'hvac'|cooling] = 0.25
```

This number is nice and positive. But not very close to one.

Now let's do heating:

```
rho = np.corrcoef(df_heating['t_out'], df_heating['hvac'])
print(f"rho['t_out', 'hvac'|heating] = {rho[0, 1]:1.2f}")
```

Output:

```
rho['t_out', 'hvac'|heating] = -0.45
```

This number is negative as expected. And it is much closer to the minimum possible value (-1) than the cooling correlation coefficient is to the maximum value $(+1)$.

Finally, let's do the off setting:

```
rho = np.corrcoef(df_off['t_out'], df_off['hvac'])
print(f"rho['t_out', 'hvac'|off] = {rho[0, 1]:1.2f}")
```

Output:

```
rho['t_out', 'hvac'|off] = -0.03
```

This is tiny and it makes sense because the HVAC system is off during off setting.

14.3 Correlation Is Not Causation

I do not think I have to stress this too much because you are engineers! But let me at least state it once:

Correlation does not imply causation.

When you have two random variables X and Y and you find that they are correlated, you cannot tell, just from this observation, that any of the following is true:

- X is causing Y,
- Y is causing X,
- X and Y have a common (unobserved) cause Z,
- or a myriad of other possibilities.

There is nothing in probability theory or statistics that can help you distinguish between these cases. You need something more. What more? Well, you need some physical knowledge.

In our example on smart buildings, we saw that the external temperature t_out and energy consumption hvac were correlated. From our understanding of the situation, we consider it much more likely that the external temperature affects the energy consumption and not the other way around!

The bottom line is that you cannot do anything useful with data science unless you augment the models you build with the causal mechanisms you (think) you know. In jargon, you have to make your causal assumptions clear. So, resist the temptation to just apply a technique to get a quick answer. Think about your problem. Write down what you know. Write down what you think is the causal mechanism. And then do data science.

It is actually beyond the scope of this class to teach you how to talk about causality. If you want to learn more, a good starting point is the following keynote lecture by Judea Pearl, a computer scientist and winner of the Turing award, on the science of causality (Pearl, 2019).

14.4   Two Uncorrelated Random Variables Are Not Necessarily Independent: Colab

We have seen that if two random variables X and Y are independent, then their covariance is zero,

$$\mathbf{C}[X, Y] = 0,$$

and therefore their correlation coefficient is also zero:

$$\rho(X, Y) = 0.$$

Does the reverse hold? Namely, if you find that the correlation between two random variables is zero, does this imply that they are independent? The answer to this question is a loud **NO**. I will show that it does not hold through a counterexample.

Take these two independent Normal random variables:

$$X \sim N(0, 1),$$

and

$$Z \sim N(0, 1).$$

Then define the new random variable Y by:

$$Y = X^2 + 0.1Z.$$

Since there is a functional relationship between the two random variables, they are obviously not independent. But let's generate some data from them and estimate the correlation:

```
xdata = np.random.randn(10000)
zdata = np.random.randn(10000)
ydata = xdata ** 2 + 0.2 * zdata
```

Figure 14.3 shows the scatter between X and Y data. Well, it's obvious that they are not independent. Let's see what the correlation coefficient is:

```
rho = np.corrcoef(xdata, ydata)
print(f"rho(X, Y) = {rho[0, 1]:1.2f}")
```

A possible output is:

```
rho(X, Y) = -0.01
```

Very close to zero. So, X and Y are uncorrelated!

After you see the scatter plot like this, you get suspicious. You start thinking that there may be a correlation between the square of X and Y. Let's estimate the correlation of X^2 and Y to see what it turns out to be:

```
rho = np.corrcoef(xdata ** 2, ydata)
print(f"rho(X^2, Y) = {rho[0, 1]:1.2f}")
```

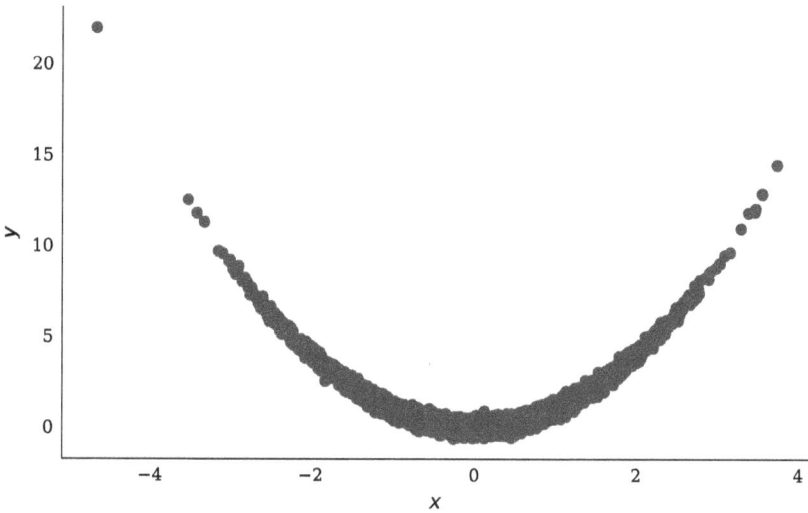

Figure 14.3. Scatter plot between X and Y data.

A possible output is:

```
rho(X^2, Y) = 0.99
```

Almost one!

Remember this example please! You cannot conclude, from just looking at the correlation coefficient, that two random variables are independent. Do the scatter plots and use your common sense. Do not just rely on a number to make decisions.

14.5   Linear Regression with One Variable: Colab

Assume that you have N pairs of observations (x_i, y_i) from two random variables X and Y. Now, you suspect that there is a linear relationship between these two random variables. Namely, you suspect that there exist coefficients a and b such that:

$$Y = aX + b + U,$$

where U is some unobserved factor that we are going to treat as noise. How can you find a and b from the N observations we have? This is a so-called *regression* problem.

Let's develop the simplest possible approach for solving the regression problem. It is called the *least squares* approach. The approach basically says that you should pick a and b so as to minimize the sum of square errors:

$$L(a, b) = \sum_{i=1}^{N} (ax_i + b - y_i)^2.$$

The error here is the prediction you are making for y_i for given a and b.

Okay, we proceed as usual when it comes to optimization. We take derivatives of L with respect to a and b, set them equal to zero and solve for a and b. Let's do it! Here is the first derivative:

$$\frac{\partial L(a, b)}{\partial a} = \frac{\partial}{\partial a} \sum_{i=1}^{N} (ax_i + b - y_i)^2$$

$$= \sum_{i=1}^{N} \frac{\partial}{\partial a} (ax_i + b - y_i)^2$$

$$= \sum_{i=1}^{N} 2(ax_i + b - y_i)x_i$$

$$= 2 \sum_{i=1}^{N} (ax_i^2 + bx_i - y_ix_i)$$

$$= 2 \left[a \sum_{i=1}^{N} x_i^2 + b \sum_{i=1}^{N} x_i - \sum_{i=1}^{N} x_iy_i \right].$$

Here is the other one:

$$\frac{\partial L(a, b)}{\partial b} = \frac{\partial}{\partial b} \sum_{i=1}^{N} (ax_i + b - y_i)^2$$

$$= \sum_{i=1}^{N} \frac{\partial}{\partial b} (ax_i + b - y_i)^2$$

$$= \sum_{i=1}^{N} 2(ax_i + b - y_i)$$

$$= 2 \sum_{i=1}^{N}(ax_i + b - y_i)$$

$$= 2 \left[a \sum_{i=1}^{N} x_i + bN - \sum_{i=1}^{N} y_i \right].$$

Setting these two derivatives to zero, yields the following linear system:

$$a \sum_{i=1}^{N} x_i^2 + b \sum_{i=1}^{N} x_i = \sum_{i=1}^{N} x_i y_i,$$

$$a \sum_{i=1}^{N} x_i + bN = \sum_{i=1}^{N} y_i.$$

This is a 2×2 system and we can solve it by substitution. Take the second equation and solve for b:

$$b = \frac{1}{N} \left[\sum_{i=1}^{N} y_i - a \sum_{i=1}^{N} x_i \right]$$

$$= \frac{1}{N} \sum_{i=1}^{N} y_i - a \frac{1}{N} \sum_{i=1}^{N} x_i$$

$$= \hat{\mu}_Y - a \hat{\mu}_X,$$

where $\hat{\mu}_X$ and $\hat{\mu}_Y$ are the empirical estimates of the mean of X and Y, respectively. Now let's take this and substitute in the first equation of the linear system. We have:

$$a \sum_{i=1}^{N} x_i^2 + (\hat{\mu}_Y - a \hat{\mu}_X) \sum_{i=1}^{N} x_i = \sum_{i=1}^{N} x_i y_i$$

$$\Rightarrow a \left[\sum_{i=1}^{N} x_i^2 - \hat{\mu}_X \sum_{i=1}^{N} x_i \right] = \sum_{i=1}^{N} x_i y_i - \hat{\mu}_Y \sum_{i=1}^{N} x_i$$

$$\Rightarrow a \frac{1}{N} \left[\sum_{i=1}^{N} x_i^2 - \hat{\mu}_X \sum_{i=1}^{N} x_i \right] = \frac{1}{N} \sum_{i=1}^{N} x_i y_i - \frac{1}{N} \hat{\mu}_Y \sum_{i=1}^{N} x_i$$

$$\Rightarrow a\left[\frac{1}{N}\sum_{i=1}^{N}x_i^2 - \hat{\mu}_X\frac{1}{N}\sum_{i=1}^{N}x_i\right] = \frac{1}{N}\sum_{i=1}^{N}x_iy_i - \hat{\mu}_Y\frac{1}{N}\sum_{i=1}^{N}x_i$$

$$\Rightarrow a\left[\frac{1}{N}\sum_{i=1}^{N}x_i^2 - \hat{\mu}_X\cdot\hat{\mu}_X\right] = \frac{1}{N}\sum_{i=1}^{N}x_iy_i - \hat{\mu}_Y\cdot\hat{\mu}_X$$

$$\Rightarrow a\left[\frac{1}{N}\sum_{i=1}^{N}x_i^2 - \hat{\mu}_X^2\right] = \frac{1}{N}\sum_{i=1}^{N}x_iy_i - \hat{\mu}_Y\cdot\hat{\mu}_X$$

Now the magic starts! Notice that the coefficient of a is actually the empirical estimate of the variance of X:

$$\hat{\sigma}_X^2 = \frac{1}{N}\sum_{i=1}^{N}(x_i - \hat{\mu}_X)^2 = \cdots = \frac{1}{N}\sum_{i=1}^{N}x_i^2 - \hat{\mu}_X^2.$$

Similarly, the right-hand side is the empirical estimate of the covariance of X and Y:

$$\hat{\sigma}_{X,Y} = \frac{1}{N}\sum_{i=1}^{N}(x_i - \hat{\mu}_X)(y_i - \hat{\mu}_Y) = \cdots = \frac{1}{N}\sum_{i=1}^{N}x_iy_i - \hat{\mu}_X\cdot\hat{\mu}_Y.$$

Putting everything together, gives us that:

$$a\hat{\sigma}_X^2 = \hat{\sigma}_{X,Y}.$$

Solving for a:

$$a = \frac{\hat{\sigma}_{X,Y}}{\hat{\sigma}_X^2} = \frac{\hat{\sigma}_{X,Y}}{\hat{\sigma}_X\hat{\sigma}_Y}\frac{\hat{\sigma}_Y}{\hat{\sigma}_X},$$

which gives a relationship between a and the correlation coefficient between X and Y:

$$\hat{a} = \hat{\rho}_{X,Y}\frac{\hat{\sigma}_Y}{\hat{\sigma}_X}.$$

This is very, very interesting. Intuitively, dividing by σ_X removes the units of X, multiplying with σ_Y puts units of Y and then you multiply with the correlation coefficient to get the direction right! Nice!

What about b? Let's substitute back into the equation:

$$\hat{b} = \hat{\mu}_Y - \hat{a}\hat{\mu}_X.$$

So, this is the difference between the mean of the data and the mean prediction of our fitted model.

Let's end the theory by looking at the final square error. We will just plug in $\hat{a}$ and $\hat{b}$ in L and see what we get. This is the result[2]:

$$L(\hat{a}, \hat{b}) = N(1 - \hat{\rho}_{X,Y}^2)\hat{\sigma}_Y^2.$$

This formula offers some insight into the error. First, notice that the error grows with the number of points. This makes sense. More points more error. Second, notice that if the correlation coefficient is zero, then the error is the biggest it can be. In this case, it is N times

---

[2]And this is the proof if you are interested:

$$L(\hat{a}, \hat{b}) = \sum_{i=1}^{N}(\hat{a}x_i + \hat{b} - y_i)^2$$

$$= \sum_{i=1}^{N}(\hat{a}x_i + \hat{\mu}_Y - \hat{a}\hat{\mu}_X - y_i)^2$$

$$= \sum_{i=1}^{N}[\hat{a}(x_i - \hat{\mu}_X) - (y_i - \hat{\mu}_Y)]^2$$

$$= \sum_{i=1}^{N}[\hat{a}^2(x_i - \hat{\mu}_X)^2 + (\hat{\mu}_Y - y_i)^2 - 2\hat{a}(x_i - \hat{\mu}_X)(\hat{\mu}_Y - y_i)]^2$$

$$= \hat{a}^2\sum_{i=1}^{N}(x_i - \hat{\mu}_X)^2 + \sum_{i=1}^{N}(\hat{\mu}_Y - y_i)^2 - 2\hat{a}\sum_{i=1}^{N}(x_i - \hat{\mu}_X)(\hat{\mu}_Y - y_i)$$

$$= \hat{a}^2 N\hat{\sigma}_X^2 + N\hat{\sigma}_Y^2 + 2\hat{a}N\hat{\sigma}_{X,Y}$$

$$= N\left[\hat{\sigma}_Y^2 + \hat{\rho}_{X,Y}^2\frac{\hat{\sigma}_Y^2}{\hat{\sigma}_X^2}\hat{\sigma}_X^2 - 2\hat{\rho}_{X,Y}\frac{\hat{\sigma}_Y}{\hat{\sigma}_X}\hat{\sigma}_{X,Y}\right]$$

$$= N\left[\hat{\sigma}_Y^2 + \hat{\rho}_{X,Y}^2\hat{\sigma}_Y^2 - 2\hat{\rho}_{X,Y}^2\hat{\sigma}_Y^2\right]$$

$$= N\left[\hat{\sigma}_Y^2 - \hat{\rho}_{X,Y}^2\hat{\sigma}_Y^2\right]$$

$$= N(1 - \hat{\rho}_{X,Y}^2)\hat{\sigma}_Y^2.$$

the variance of Y. Third, the error becomes zero if the correlation of coefficient is exactly one or minus one. This only happens when the model is perfectly linear.

14.5.1   *Example: Modeling* hvac *as a function of* t_out

Let's now apply what we learned in the previous section to the smart buildings dataset. In particular, we are going to estimate the linear relationship between $X = $ t_out and $Y = $ hvac for heating and cooling.

We start with cooling.

```
xdata = df_cooling['t_out']
ydata = df_cooling['hvac']

# Estimate the statistics
mu_X = xdata.mean()
mu_Y = ydata.mean()
sigma_X = xdata.std()
sigma_Y = ydata.std()
rho_XY = np.corrcoef(xdata, ydata)[0, 1]

# The a coefficient
a = rho_XY * sigma_Y / sigma_X
# The b coefficient
b = mu_Y - a * mu_X

# Print the results
print('Model coefficients for cooling: ')
print(f'a = {a:1.2f}')
print(f'b = {b:1.2f}')
```

Output:

```
Model coefficients for cooling:
a = 3.78
b = -229.95
```

Let's evaluate the model at some points:

```
xs = np.linspace(xdata.min(), xdata.max())
ys = a * xs + b
```

Figure 14.4 shows the model and the data.

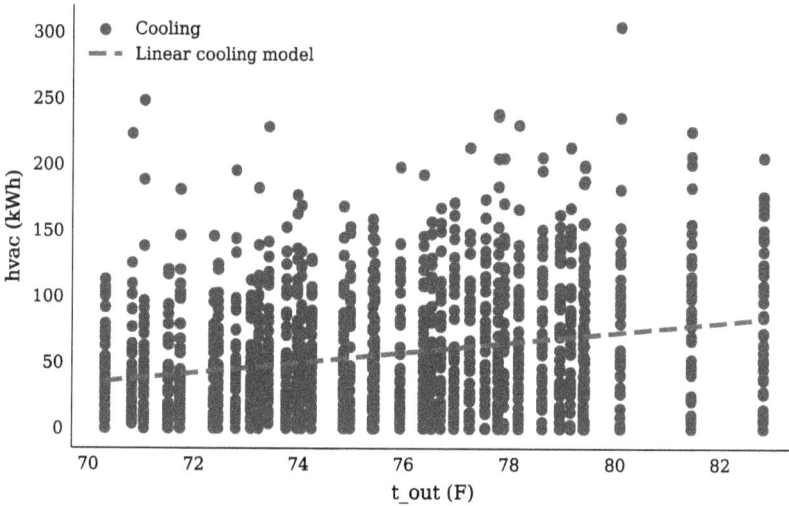

Figure 14.4.   Model and data for cooling.

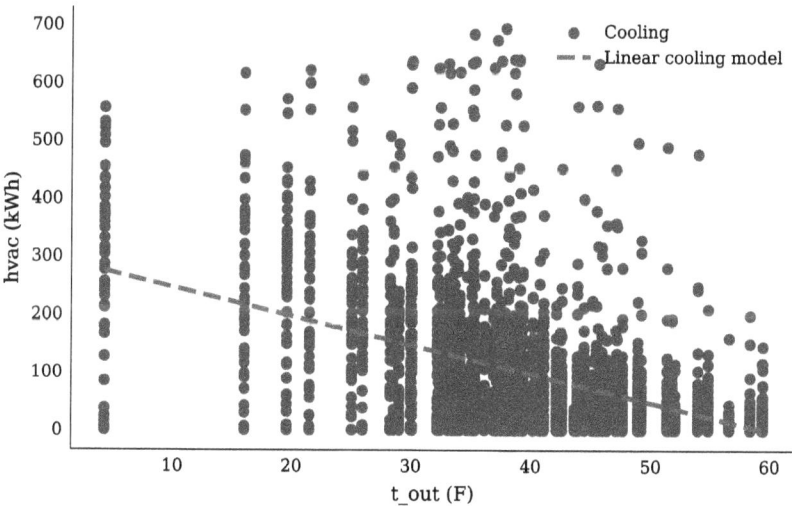

Figure 14.5.   Model and data for heating.

For heating, we can do the same. Figure 14.5 shows the corresponding model and data.

Exercises

1. **Joint probability mass function, covariance, and correlation**

   In this problem, you will perform some analytical calculations that demonstrate further the concept of the joint probability mass (density) function.

   Consider two random variables X and Y taking discrete values in $\{0, 1\}$. The joint probability mass function $p(x, y)$ is given by:

   $$p(X = 0, Y = 0) = 0.1,$$
   $$p(X = 1, Y = 0) = 0.2,$$
   $$p(X = 0, Y = 1) = 0.2,$$

   and

   $$p(X = 1, Y = 1) = 0.5.$$

   Put these numbers in a 2×2 matrix:

   ```
   A = np.array([[0.1, 0.2],
                 [0.2, 0.5]])
   ```

   and answer the following questions with Python code.

   (a) Verify that the sum of all entries in "A" is one, i.e., verify that $\sum_{x,y} p(x, y) = 1$.

   (b) What is the probability mass function of Y? *Hint*: Use the marginalization property,

   $$p(Y = y) = \sum_{x} p(x, Y = y) = p(X = 0, Y = y)$$
   $$+ p(X = 1, Y = y),$$

   to find $p(Y = y)$ for $y = 0$ and 1.

   (c) What is the probability mass function of X? *Hint*: Same as above if you swap x and y.

   (d) What is the expectation of X?

   (e) What is the expectation of Y?

(f) What is the variance of X?

(g) What is the covariance of X and Y?

(h) What is the correlation of X and Y?

(i) What is the covariance of X with itself? *Hint*: Use the definition of covariance.

(j) What is the covariance of $3X$ with Y? *Hint*: Use the definition of covariance.

(k) What is the covariance of X with 0.31? Yes, the constant 0.31. *Hint*: Use the definition of covariance for $Y = 0.31$. You will need the mean of the constant. What is the mean of a constant?

2. **Estimating the mechanical properties of a plastic material from molecular dynamics simulations**
First, download the dataset `stress_strain.txt` from the book's GitHub repository. Load it into a pandas dataframe:

```
data =  pd.read_csv('stress_strain.txt', delimiter='\t',
↳  names=['Strain', 'Stress'], skiprows=1)
data.round(3)
```

	Strain	Stress
0	0.000	-29.490
1	0.000	11.780
2	0.001	45.195
3	0.001	49.495
4	0.001	42.283
...	...	...
996	0.251	115.897
997	0.251	125.219
998	0.251	118.189
999	0.251	142.209
1000	0.252	153.354

1001 rows × 2 columns

The dataset was generated using a molecular dynamics simulation of a plastic material (many thanks to Professor Alejandro Strachan for sharing the data!).[3] This is a simulated version of the real

---

[3] In molecular dynamics, you basically simulate the motion of atoms in a material, say a metal, by solving Newton's law of motion for each atom.

experiment we saw in the Exercises of 5. Specifically, Strachan's group did the following:

- They took a rectangular chunk of the material and marked the position of each one of its atoms.
- They started applying a tensile force along one dimension. The atoms are coupled together through electromagnetic forces and they must all satisfy Newton's law of motion.
- For each value of the applied tensile force, they marked the Stress (force be unit area) in the middle of the material and the corresponding Strain of the material (percent elongation in the pulling direction).
- Eventually, the material entered the plastic regime and then it broke.

Figure 14.6 shows the data. Note that for each particular value of the strain, you don't necessarily get a unique stress. This is because, in molecular dynamics, the atoms are jiggling around due to thermal effects. So there is always this "jiggling" noise when you are trying to measure the stress and the strain. We would like to process this noise in order to extract what is known as the *stress-strain curve* of the material. The stress-strain curve

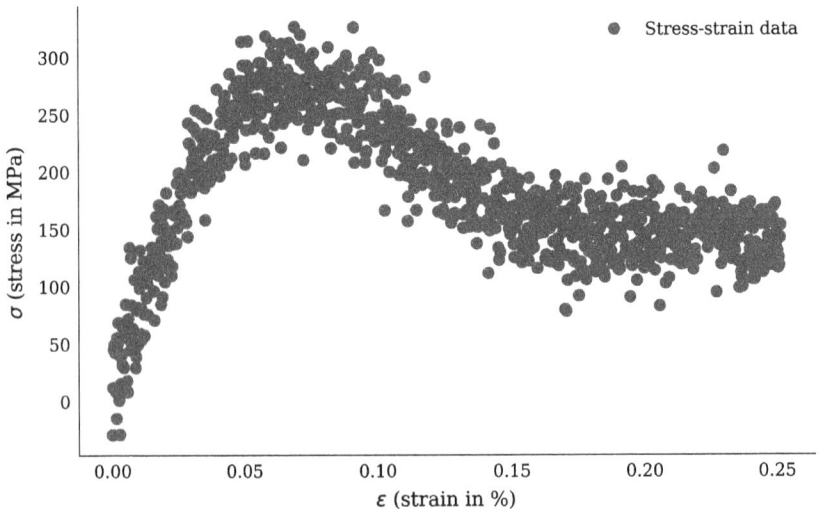

Figure 14.6. Stress-strain data.

is a macroscopic property of the material which is affected by the fine structure, e.g., the chemical bonds, the crystalline structure, any defects, etc. It is a required input to mechanics of materials. The very first part of the stress-strain curve should be linear. It is called the *elastic regime*. In that region, say $\epsilon < \epsilon_l = 0.04$, the relationship between stress and strain is:

$$\sigma(\epsilon) = E\epsilon.$$

The constant E is known as the *Young modulus* of the material. Assume that you measure ϵ without any noise, but your measured σ is noisy.

First, extract the relevant data for this problem, split it into training and validation datasets, and visualize the training and validation datasets using different colors.

```
# The point at which the stress-strain curve stops being linear
epsilon_l = 0.04
# Relevant data (this is nice way to get the linear part of the
↪   stresses and straints)
data_linear = data[data['Strain'] < 0.04]
data_nonlinear = data[data['Strain'] >= 0.04]
```

Figure 14.7 shows the linear and the non-linear part of the data with different symbols. Answer the following questions:

(a) Find the empirical correlation between Strain and Stress in the linear regime.
(b) Find the empirical correlation between Strain and Stress in the nonlinear regime.
(c) Find the empirical mean of the Strain in the linear regime.
(d) Find the standard deviation of the Strain in the linear regime.
(e) Find the empirical mean of the Stress in the linear regime.
(f) Find the standard deviation of the Stress in the linear regime.
(g) In the linear regime, fit the following model:

$$\sigma(\epsilon) = a\epsilon + b.$$

In this model, $\hat{a}$ will be your estimate of the Young's modulus E of the material. *Hint*: Use the formulas for $\hat{a}$ and $\hat{b}$ that we derived in 14.5.

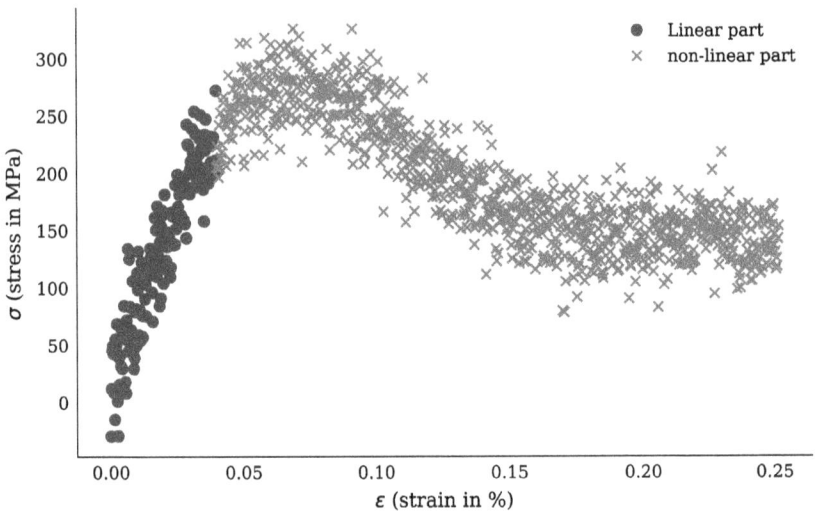

Figure 14.7.   Linear and non-linear part of the data.

(h) Theoretically, b should be exactly zero. If you get a non-zero value, where do you think this is coming from?

(i) In the same figure, plot the data from the linear regime along with the predictions of the fitted model $\sigma(\epsilon) = \hat{a}\epsilon + \hat{b}$.

Chapter 15

Linear Regression

In this chapter, I cast the least squares problem in the language of linear algebra. I introduce polynomial regression and the generalized linear model. I demonstrate the concepts of overfitting and underfitting, and I explain how to use cross validation to select the appropriate number of basis functions. I end by discussing the maximum likelihood interpretation of least squares and use it to estimate the variance of the measurement noise.

15.1 Regression with One Variable Revisited

Let's say that we have N observations consisting of inputs (features):

$$x_{1:N} = (x_1, \ldots, x_N),$$

and outputs (targets):

$$y_{1:N} = (y_1, \ldots, y_N).$$

We want to learn the map (function) that connects the inputs to the outputs. This type of problem is called *supervised learning* because we have a teacher (the data) that tells us what the correct output is for each input. We say that we have a *regression* problem when the outputs are continuous quantities, e.g., mass, height, price. When the outputs are discrete, e.g., colors, numbers, then we say that we have a *classification* problem. In this chapter, the focus is on regression.

To proceed, we need to make a model that connects the inputs to the outputs. The simplest such model is:

$$y = w_0 + w_1 x + \text{measurement noise.}$$

This is the linear model we saw in the previous chapter with $w_0 = b$ and $w_1 = a$. The parameters w_0 and w_1 are called the regression *weights* and we need to find them using the observations $x_{1:N}$ and $y_{1:N}$.

In the previous chapter, we fitted the model by minimizing the sum of square errors:

$$L(w_0, w_1) = \sum_{i=1}^{N} (y_i - w_0 - w_1 x_i)^2.$$

Now we are going to express this equation using linear algebra. We do this for two reasons. First, it is a lot of fun! Second, it is essential for formulating the fitting problem for more complicated models.

We will need the *design matrix*:

$$\mathbf{X} = \begin{bmatrix} 1 & x_1 \\ 1 & x_2 \\ \vdots & \vdots \\ 1 & x_N \end{bmatrix}.$$

The design matrix $\mathbf{X}$ is a $N \times 2$ matrix with the first column being just one and the second column being the observed inputs. We will also need, the *vector of observed outputs*:

$$\mathbf{y} = y_{1:N} = (y_1, \ldots, y_N),$$

and the *vector of weights*:

$$\mathbf{w} = (w_0, w_1).$$

I hope that you know how to do matrix-vector multiplication. Notice what we get when we multiply the design matrix $\mathbf{X}$ with the weight vector $\mathbf{w}$:

$$
\mathbf{Xw} = \begin{bmatrix} 1 & x_1 \\ 1 & x_2 \\ \vdots & \vdots \\ 1 & x_N \end{bmatrix} \cdot \begin{bmatrix} w_0 \\ w_1 \end{bmatrix}
$$

$$
= \begin{bmatrix} w_0 + w_1 x_1 \\ w_0 + w_1 x_2 \\ \vdots \\ w_0 + w_1 x_N \end{bmatrix}.
$$

Wow! So, $\mathbf{Xw}$ is an N-dimensional vector that contains the predictions of our linear model at the observed inputs. If we subtract this vector from the vector of observed outputs $\mathbf{y}$, we get the prediction errors:

$$
\mathbf{y} - \mathbf{Xw} = \begin{bmatrix} y_1 - w_0 - w_1 x_1 \\ y_2 - w_0 - w_1 x_2 \\ \vdots \\ y_N - w_0 - w_1 x_N \end{bmatrix}.
$$

Now recall that the *Euclidean norm* $\|\mathbf{v}\|$ of a vector is the square root of the sum of the squares of its components. Let's take the square of the Euclidean norm of the error vector. It is:

$$
\|\mathbf{y} - \mathbf{Xw}\|^2 = \sum_{i=1}^{N} (y_i - w_0 - w_1 x_i)^2.
$$

But this is just sum of square errors, i.e., we have shown that:

$$
L(w_0, w_1) = L(\mathbf{w}) = \|\mathbf{y} - \mathbf{Xw}\|^2.
$$

We have managed to express the loss function using linear algebra. The mathematical problem of finding the best weight vector is now:

$$
\min_{\mathbf{w}} L(\mathbf{w}) = \min_{\mathbf{w}} \|\mathbf{y} - \mathbf{Xw}\|^2.
$$

This form is much more convenient mathematically.

Remember that to solve the minimization problem, we need to take the gradient of $L(\mathbf{w})$ with respect to $\mathbf{w}$ and set the result equal to zero. Expressing the loss function in this way allows us to take derivatives in a much easier way. But there is one more thing that we could do before we take the gradient. Notice that the Euclidean norm of a vector $\mathbf{v}$ satisfies:

$$\|\mathbf{v}\|^2 = \mathbf{v}^T\mathbf{v},$$

where we are thinking of $\mathbf{v}$ as a column matrix and $\mathbf{v}^T$ is the transpose of $\mathbf{v}$, i.e., a row matrix. To prove the equality, we start from the right-hand side:

$$\mathbf{v}^T\mathbf{v} = \begin{bmatrix} v_1 & v_2 & \cdots & v_N \end{bmatrix} \cdot \begin{bmatrix} v_1 \\ v_2 \\ \vdots \\ v_N \end{bmatrix}$$

$$= \sum_{i=1}^{N} v_i^2$$

$$= \|\mathbf{v}\|^2.$$

Interesting! So we can rewrite the sum of square errors as:

$$L(\mathbf{w}) = \|\mathbf{y} - \mathbf{Xw}\|^2$$

$$= (\mathbf{y} - \mathbf{Xw})^T (\mathbf{y} - \mathbf{Xw})$$

$$= \left[\mathbf{y}^T - (\mathbf{Xw})^T \right] (\mathbf{y} - \mathbf{Xw})$$

$$= \left(\mathbf{y}^T - \mathbf{w}^T\mathbf{X}^T \right) (\mathbf{y} - \mathbf{Xw})$$

$$= \mathbf{y}^T\mathbf{y} - \mathbf{w}^T\mathbf{X}^T\mathbf{y} - \mathbf{y}^T\mathbf{Xw} + \mathbf{w}^T\mathbf{X}^T\mathbf{Xw}$$

Now, because $\mathbf{w}^T\mathbf{X}^T\mathbf{y}$ is just a number (think about the dimensions $(1 \times 2) \times (2 \times N) \times (N \times 1) = 1 \times 1$), it is the same as its transpose, i.e.,

$$\mathbf{w}^T\mathbf{X}^T\mathbf{y} = \left(\mathbf{w}^T\mathbf{X}^T\mathbf{y} \right)^T = \mathbf{y}^T\mathbf{Xw}.$$

Using this fact, we can write:

$$L(\mathbf{w}) = \mathbf{y}^T\mathbf{y} - 2\mathbf{w}^T\mathbf{X}^T\mathbf{y} + \mathbf{w}^T\mathbf{X}^T\mathbf{X}\mathbf{w}.$$

Now we can take the gradient with respect to $\mathbf{w}$:

$$\nabla_{\mathbf{w}}L(\mathbf{w}) = \nabla_{\mathbf{w}}\left[\mathbf{y}^T\mathbf{y} - 2\mathbf{w}^T\mathbf{X}^T\mathbf{y} + \mathbf{w}^T\mathbf{X}^T\mathbf{X}\mathbf{w}\right]$$
$$= -2\mathbf{X}^T\mathbf{y} + 2\mathbf{X}^T\mathbf{X}\mathbf{w}.$$

Okay, I do admit that I did some derivative magic there. But the result is correct. If you really want to understand it, you would have to work out the gradient of the following $\mathbf{w}^T\mathbf{u}$ and $\mathbf{w}^T\mathbf{A}\mathbf{w}$ where $\mathbf{u}$ is a 2-dimensional vector and $\mathbf{A}$ is a 2×2 matrix.

Setting the gradient to zero, yields the following linear system:

$$\mathbf{X}^T\mathbf{X}\mathbf{w} = \mathbf{X}^T\mathbf{y}.$$

The bottom line: to find the best $\mathbf{w}$, you must solve this linear system! As I will show you in a while, for more complex models that remain linear in the parameters you basically have to do exactly the same thing but with a different design matrix. By the way, if you work out the analytical solution for a linear model with 2 parameters you will get exactly the expression with the correlation between X and Y we derived in the previous chapter.

15.2 Example: Linear Regression with a Single Variable: Colab

Let's create a *synthetic* dataset to introduce the basic concepts. By synthetic, I mean that we know the ground truth and we can use it to test the performance of the model. Let's start with pairs of x and y which definitely have a linear relationship, albeit y may be contaminated with Gaussian noise. In particular, we generate the data from:

$$y_i = -0.5 + 2x_i + 0.1\epsilon_i,$$

where $\epsilon_i \sim N(0,1)$ and where we sample $x_i \sim U([0,1])$. Here is how to generate this synthetic dataset:

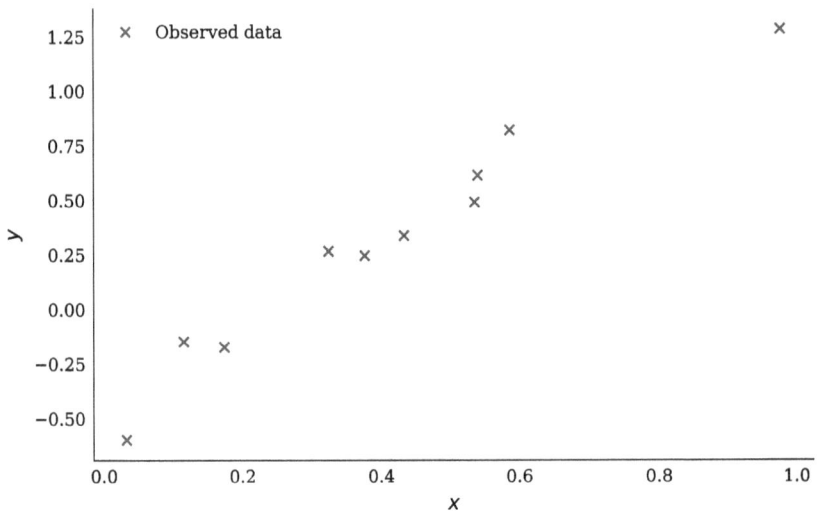

Figure 15.1.　Synthetic data for linear regression with a single variable.

```
num_obs = 10
x = np.random.rand(num_obs)
w0_true = -0.5
w1_true = 2.0
sigma_true = 0.1
y = w0_true + w1_true * x + sigma_true * np.random.randn(num_obs)
```

Figure 15.1 shows the data.

　　We will now use least squares to fit the data to this linear model:

$$y = w_0 + w_1 x.$$

As we discussed in the previous section, least squares minimize the square loss:

$$L(\mathbf{w}) = \sum_{i=1}^{N}(y_i - w_0 - w_1 x_i)^2 = \|\mathbf{y} - \mathbf{X}\mathbf{w}\|^2,$$

where $\mathbf{y} = (y_1, \ldots, y_N)$ is the vector of observations, $\mathbf{w} = (w_0, w_1)$ is the weight vector, and the $N \times 2$ design matrix $\mathbf{X}$ is:

$$\mathbf{X} = \begin{bmatrix} 1 & x_1 \\ 1 & x_2 \\ \vdots & \vdots \\ 1 & x_N \end{bmatrix}.$$

We definitely need to make the design matrix $\mathbf{X}$:

```
X = np.hstack([np.ones((num_obs, 1)), x.reshape((num_obs, 1))])
X
```

A possible output:

```
array([[1.        , 0.42008217],
       [1.        , 0.40615312],
       [1.        , 0.49311771],
       [1.        , 0.98130335],
       [1.        , 0.45972574],
       [1.        , 0.36060826],
       [1.        , 0.53511177],
       [1.        , 0.18104067],
       [1.        , 0.6172522 ],
       [1.        , 0.1663081 ]])
```

Once we have this matrix, we can use `numpy.linalg.lstsq` to solve the least squares problem. This function solves in a smart way the linear system we derived in the previous section, i.e.,

$$\mathbf{X}^T \mathbf{X} \mathbf{w} = \mathbf{X}^T \mathbf{y}.$$

It works as follows:

```
w, _, _, _ = np.linalg.lstsq(X, y, rcond=None)
print(f'w_0 = {w[0]:1.2f}')
print(f'w_1 = {w[1]:1.2f}')
```

A possible output:

```
w_0 = -0.51
w_1 = 2.06
```

So, you see that the values we found for w_0 and w_1 are close to the correct values, but not exactly the same. That is fine. There is noise in the data and we have only used ten observations. The more noise there is, the more observations it would take to identify the regression coefficients correctly.

Let's now plot the regression function against the data:

```
fig, ax = make_full_width_fig()
# Some points on which to evaluate the regression function
xx = np.linspace(0, 1, 100)
# The true connection between x and y
yy_true = w0_true + w1_true * xx
# The model we just fitted
yy = w[0] + w[1] * xx
# plot the data again
ax.plot(x, y, 'x', label='Observed data')
# overlay the true
ax.plot(xx, yy_true, label='True response surface')
# overlay our prediction
ax.plot(xx, yy, '--', label='Fitted model')
plt.legend(loc='best')
```

Figure 15.2 shows the result.

15.2.1   An example where things do not work as expected: Underfitting

Let's try to fit a linear regression model to data generated from:

$$y_i = -0.5 + 2x_i + 2x_i^2 + \epsilon_i,$$

where $\epsilon_i \sim N(0,1)$ and where we sample $x_i \sim U([-1,1])$:

```
num_obs = 10
x = -1.0 + 2 * np.random.rand(num_obs)
w0_true = -0.5
w1_true = 2.0
```

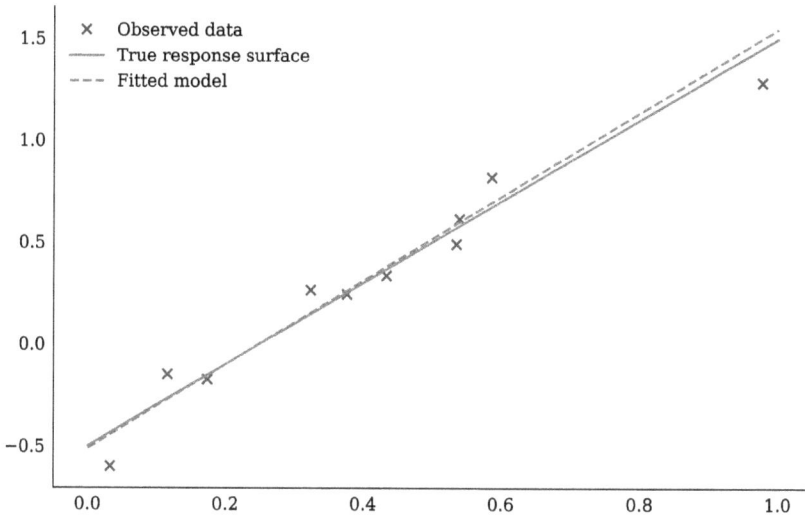

Figure 15.2. Linear regression function against the data from linear ground truth.

```
w2_true = 2.0
sigma_true = 0.1
y = w0_true + w1_true * x + w2_true * x ** 2 + sigma_true *
↪    np.random.randn(num_obs)
```

We will still try to fit a linear model to this dataset. We know that it is not going to work well, but let's try it anyway. First, create the design matrix just like before:

```
X = np.hstack([np.ones((num_obs, 1)), x.reshape((num_obs, 1))])
w, _, _, _ = np.linalg.lstsq(X, y, rcond=None)
print(f'w_0 = {w[0]:1.2f}')
print(f'w_1 = {w[1]:1.2f}')
```

A possible output:

```
w_0 = 0.13
w_1 = 2.16
```

Figure 15.3 shows the result. Notice that the model is too simple to fit the data. This is an example of *underfitting*. How can we fix this?

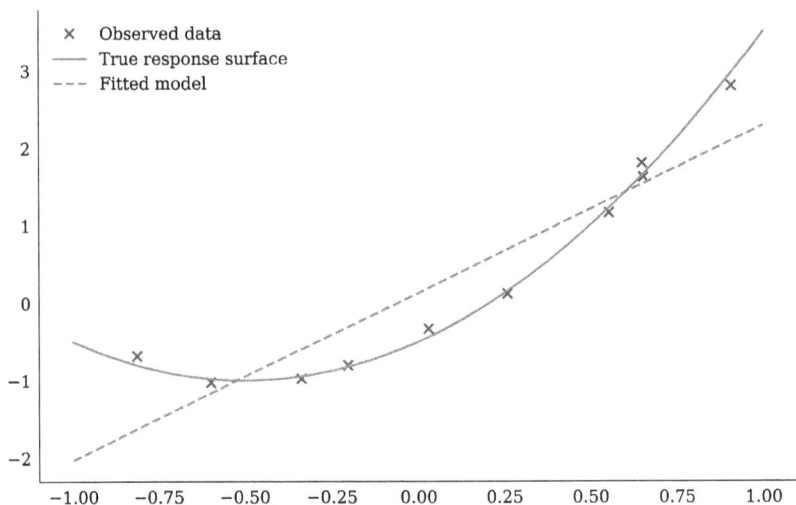

Figure 15.3. Linear regression function against the data from the quadratic ground truth. An example of underfitting.

15.3 Polynomial Regression: Colab

In the previous section, we tried to fit a linear regression model to data generated from a quadratic relationship. The linear model underfit the data.

We want to try to fit a quadratic model:

$$y = w_0 + w_1 x + w_2 x^2.$$

How can we do this? Of course, by minimizing the square loss:

$$L(w_0, w_1, w_2) = \sum_{i=1}^{N} (y_i - w_0 - w_1 x_i - w_2 x_i^2)^2.$$

Fortunately, we do not have to do things from scratch. Now it is time for the linear algebra we developed to pay off. Recall that $\mathbf{y} = (y_1, \ldots, y_N)$ is the vector of observations. Use

$$\mathbf{w} = (w_0, w_1, w_2),$$

to denote the weight vector.

What about the design matrix? Before, it was an $N \times 2$ matrix with the first column being one and the second column being the vector of observed inputs. Now it is the $N \times 3$ matrix. The first two columns are exactly like before, but now the third column is the observed inputs squared. So, it is:

$$\mathbf{X} = \begin{bmatrix} 1 & x_1 & x_1^2 \\ 1 & x_2 & x_2^2 \\ \vdots & \vdots & \\ 1 & x_N & x_N^2 \end{bmatrix}.$$

As before, if you multiply the design matrix $\mathbf{X}$ with the weight vector $\mathbf{w}$ you get the predictions of our model. So, again, the square loss can be written as:

$$L(w_0, w_1, w_2) = L(\mathbf{w}) = \|\mathbf{y} - \mathbf{Xw}\|^2.$$

Well, this is mathematically the same equation as before. The only difference is that we have 3-dimensional weight vector (instead of a 2-dimensional) and that the design matrix is $N \times 3$ instead of $N \times 2$. If you take the gradient of this with respect to $\mathbf{w}$ and set it equal to zero you will get that you need to solve exactly the same linear system of equations as before (but now it is 3 equations for 3 unknowns instead of 2 equations for 2 unknowns).

Let's solve it numerically. First, the design matrix:

```
X = np.hstack([np.ones((num_obs, 1)), x.reshape((num_obs, 1)),
↪   x.reshape((num_obs, 1)) ** 2])
```

and then:

```
w, _, _, _ = np.linalg.lstsq(X, y, rcond=None)
print(f'w_0 = {w[0]:1.2f}')
print(f'w_1 = {w[1]:1.2f}')
print(f'w_2 = {w[2]:1.2f}')
```

A possible output:

```
w_0 = -0.54
w_1 = 1.98
w_2 = 2.06
```

Notice that we have almost discovered the true parameters. Figure 15.4 shows the result.

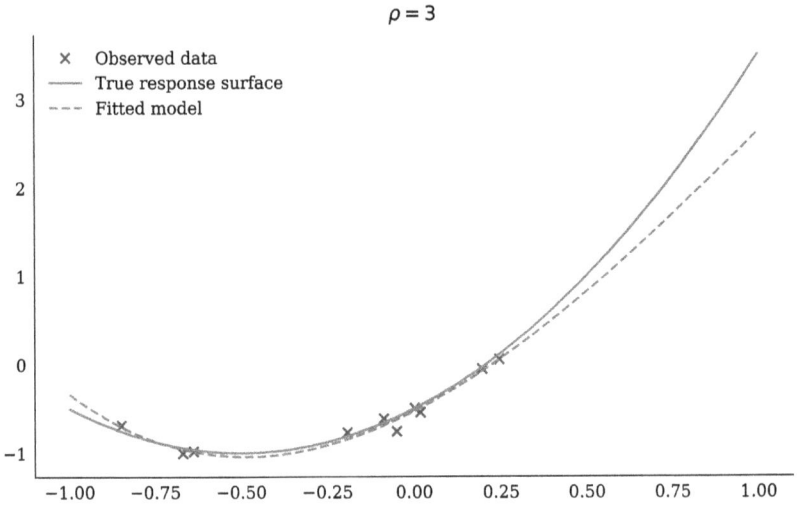

Figure 15.4. Quadratic regression against the data from quadratic ground truth.

15.3.1 *Regression with high-degree polynomials and overfitting*

What would have happened if we tried to fit a higher degree polynomial. To achieve this, we need to be able to evaluate a design matrix of the form:

$$
\mathbf{X} = \begin{bmatrix}
1 & x_1 & x_1^2 \cdots & x_1^\rho \\
1 & x_2 & x_2^2 \cdots & x_2^\rho \\
\vdots & \vdots & \vdots \cdots & \vdots \\
1 & x_N & x_N^2 \cdots & x_N^\rho
\end{bmatrix},
$$

where ρ is the degree of the polynomial. The linear system we need to solve is the same as before. Only the weight vector is now $\rho + 1$ dimensional and the design matrix $N \times (\rho + 1)$. Let's write some code to find the design matrix:

```
def get_polynomial_design_matrix(x, degree):
    """Returns the polynomial design matrix of ``degree`` evaluated
↪   at ``x``.
    """

    # Make sure this is a 2D numpy array with only one column
```

```
assert isinstance(x, np.ndarray), 'x is not a numpy array.'
assert x.ndim == 2, 'You must make x a 2D array.'
assert x.shape[1] == 1, 'x must be a column.'
# Start with an empty list where we are going to put the columns
↪  of the matrix
cols = []
# Loop over columns and add the polynomial
for i in range(degree+1):
    cols.append(x ** i)
return np.hstack(cols)
```

Let's try fitting a degree 3 polynomial and see what we get:

```
degree = 3
X = get_polynomial_design_matrix(x[:, None], degree)
w, _, _, _ = np.linalg.lstsq(X, y, rcond=None)
print('w = ', w)
```

A possible output:

```
w = [-0.52877742  1.99159421  1.67155185 -0.50839761]
```

Let's make predictions. Notice that for making predictions I am evaluating the design matrix on the points I want to make predictions at:

```
xx = np.linspace(-1, 1, 100)
# The true connection between x and y
yy_true = w0_true + w1_true * xx + w2_true * xx ** 2
# The model we just fitted
XX = get_polynomial_design_matrix(xx[:, None], degree)
yy = np.dot(XX, w)
```

Figure 15.5 shows the result.

Now let's repeat the same exercise for a degree 4 polynomial. The code is exactly the same as before, except that we change the degree to 4. Figure 15.6 shows the result. Notice that the regression fits the data, but it deteriorates outside the range of the data. This is an example of *overfitting*. There is a balance between underfitting and overfitting. The more complex the model, the more likely it is to overfit. We always strive to find the simplest model that fits the data well.

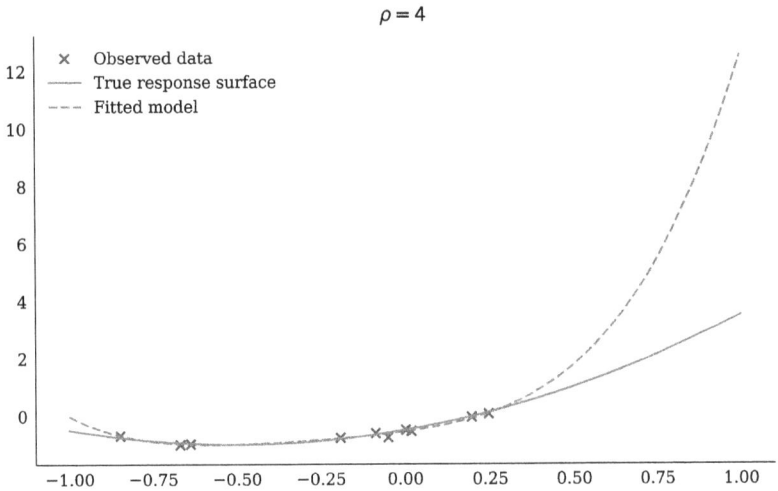

Figure 15.5. Degree 3 polynomial regression against the data from quadratic ground truth.

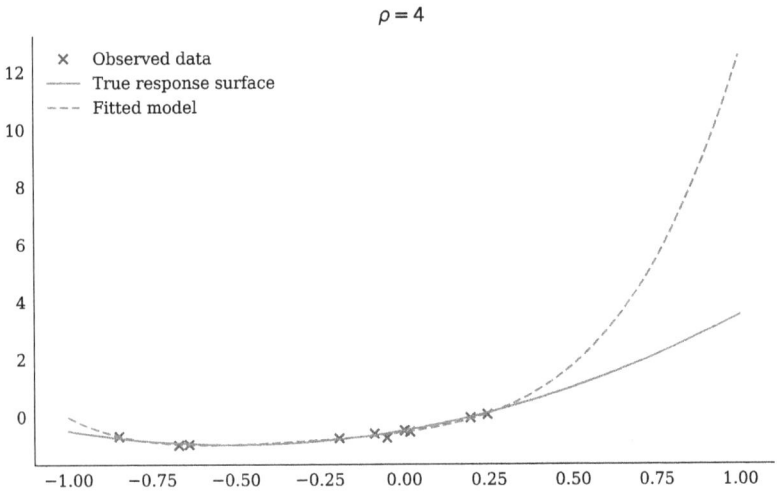

Figure 15.6. Degree 4 polynomial regression against the data from quadratic ground truth. An example of overfitting.

15.4   The Generalized Linear Model: Colab

Let me now show you the most general form of a linear model. It is called the *generalize linear model*.

The form of the generalized linear model is:

$$y(\mathbf{x}; \mathbf{w}) = \sum_{j=1}^{m} w_j \phi_j(\mathbf{x}) = \mathbf{w}^T \boldsymbol{\phi}(\mathbf{x})$$

where the weight vector is:

$$\mathbf{w} = (w_1, \ldots, w_m)^T$$

and

$$\boldsymbol{\phi} = (\phi_1, \ldots, \phi_m)^T$$

are arbitrary *basis functions*. Note that the model is linear in the parameters $\mathbf{w}$, but the basis functions $\boldsymbol{\phi}(\mathbf{x})$ can be non-linear.

15.4.1 *The polynomial model as a generalized linear model*

We have already seen an example of a generalized linear model when $\mathbf{x}$ has only one dimension: the polynomial model. In the polynomial model, the basis functions are:

$$\phi_1(x) = 1,$$
$$\phi_2(x) = x,$$
$$\phi_3(x) = x^2,$$

and so on.

15.4.2 *Multivariate linear regression as a generalized linear model*

In multivariate linear regression, the inputs $\mathbf{x}$ have d dimensions, say

$$\mathbf{x} = (x_1, \ldots, x_d).$$

The linear model is:

$$y = w_0 + w_1 x_1 + w_2 x_2 + \cdots w_d x_d.$$

This is also a generalized linear model with $m = d + 1$ basis functions:

$$\phi_1(\mathbf{x}) = 1,$$
$$\phi_2(\mathbf{x}) = x_1,$$
$$\phi_3(\mathbf{x}) = x_2,$$

and so on.

15.4.3 Other generalized linear models

Some common examples of generalized linear models include:

- Fourier series are defined on a domain $[0, L]$ and have basis functions:

$$\phi_{2j}(x) = \cos\left(\frac{2j\pi}{L}x\right) \quad \text{and} \quad \phi_{2j+1}(x) = \sin\left(\frac{2j\pi}{L}x\right),$$

for $j = 1, \ldots, m/2$. The Fourier features are excellent for representing functions that have some periodic or semi-periodic structure.
- Radial basis functions can be defined over an arbitrary domain Ω and have basis functions:

$$\phi_j(\mathbf{x}) = \exp\left\{-\frac{\|\mathbf{x} - \mathbf{x}_j\|^2}{2\ell^2}\right\},$$

where $\mathbf{x}_j$ is the center of the jth basis function and ℓ is the width of the basis function. The radial basis functions are excellent for representing functions that have a localized structure.

15.4.4 Fitting the generalized linear model using least squares

The idea is to find the best $\mathbf{w}$ by minimizing a quadratic loss function:

$$\mathcal{L}(\mathbf{w}) = \sum_{i=1}^{N} [y(\mathbf{x}_i; \mathbf{w}) - y_i]^2.$$

As we discussed in the previous sections, the loss function can be re-expressed as:

$$\mathcal{L}(\mathbf{w}) = \|\mathbf{\Phi}\mathbf{w} - \mathbf{y}\|^2$$
$$= (\mathbf{\Phi}\mathbf{w} - \mathbf{y})^T (\mathbf{\Phi}\mathbf{w} - \mathbf{y}).$$

Here $\mathbf{\Phi} \in \mathbb{R}^{n \times m}$ is the design matrix:

$$\Phi_{ij} = \phi_j(\mathbf{x}_j).$$

So the design matrix is $N \times M$ where N is the number of observations and M is the number of basis functions. Furthermore, the ith column of the design matrix is the ith basis function evaluated at all N observed inputs.

To minimize the loss function, we follow these steps. First, take the derivative of $\mathcal{L}(\mathbf{w})$ with respect to $\mathbf{w}$. Then, set it equal to zero and solve for $\mathbf{w}$. You will get, see (Bishop, 2006), the following linear system:

$$\left(\mathbf{\Phi}^T \mathbf{\Phi}\right) \mathbf{w} = \mathbf{\Phi}^T \mathbf{y}.$$

This is mathematically identical to what we had for the linear and polynomial regression! The only difference is that we now call the design matrix $\mathbf{\Phi}$ instead of $\mathbf{X}$.

To solve this problem, just use `numpy.linalg.lstsq`. You give it $\mathbf{\Phi}$ and $\mathbf{y}$ and it returns the $\mathbf{w}$ that solves the linear system.

15.4.5 *Example: Motorcycle data with polynomials*

In this example, we will use the famous motorcycle dataset (Silverman, 1985). The data come from a simulated motorcycle crash experiment and they include accelerometer readings at the helmet of the motorcycle driver at various time points, see Figure 15.7. You can download the data from the book's website. We will perform regression between time and the acceleration.

Let's start with polynomial regression. Figure 15.8 shows the result. Notice that the model is too simple to fit the data. Now, if you experiment with higher degree polynomials, you will see that there is not a good fit for any polynomial degree. We kind of pass immediately from underfitting to overfitting. We need a completely different model to fit this dataset well.

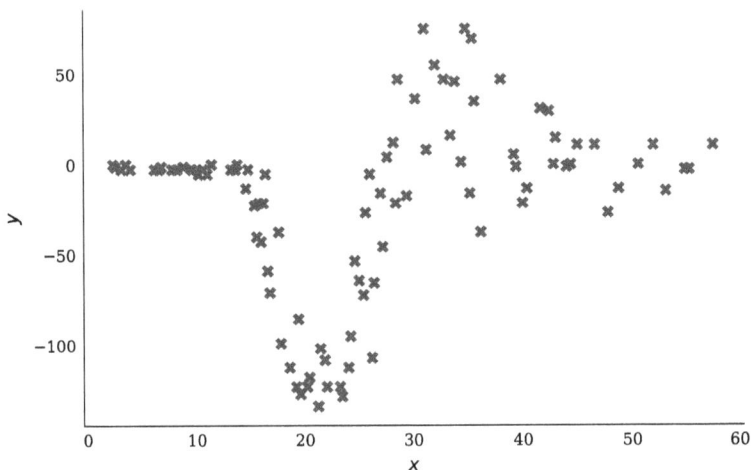

Figure 15.7. Motorcycle data.

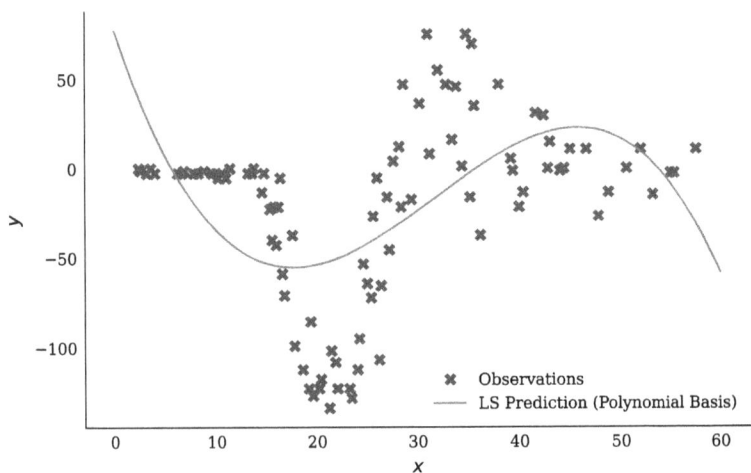

Figure 15.8. Polynomial regression.

15.4.6　*Example: Motorcycle data with radial basis functions*

Let's now repeat what we did with polynomial regression with a Fourier basis. The mathematical form of the basis is:

$$\phi_{2j}(x) = \cos\left(\frac{2j\pi}{L}x\right),$$

and

$$\phi_{2j+1}(x) = \sin\left(\frac{2j\pi}{L}x\right),$$

for $j = 1, \ldots, m/2$. First, we write code that computes the design matrix for the new basis:

```
def get_fourier_design_matrix(x, L, num_terms):
    """Fourier expansion with ``num_terms`` cosines and sines.

    Arguments:

        L              -      The "length" of the domain.
        num_terms      -      How many Fourier terms do you want. This
                              is not the number
                              of basis functions you get. The number
                              of basis functions
                              is 1 + num_terms / 2. The first one is a
                              constant.
    """
    # Make sure this is a 2D numpy array with only one column
    assert isinstance(x, np.ndarray), 'x is not a numpy array.'
    assert x.ndim == 2, 'You must make x a 2D array.'
    assert x.shape[1] == 1, 'x must be a column.'
    N = x.shape[0]
    cols = [np.ones((N, 1))]
    # Loop over columns and add the polynomial
    for i in range(int(num_terms / 2)):
        cols.append(np.cos(2 * (i+1) * np.pi / L * x))
        cols.append(np.sin(2 * (i+1) * np.pi / L * x))
    return np.hstack(cols)
```

Let's start by visualizing the Fourier basis:

```
fig, ax = plt.subplots()
xx = np.linspace(0, 60, 200)
Phi_xx = get_fourier_design_matrix(xx[:, None], 60.0, 4)
plt.plot(xx, Phi_xx)
plt.ylabel(r'$\phi_i(x)$')
plt.xlabel('$x$')
```

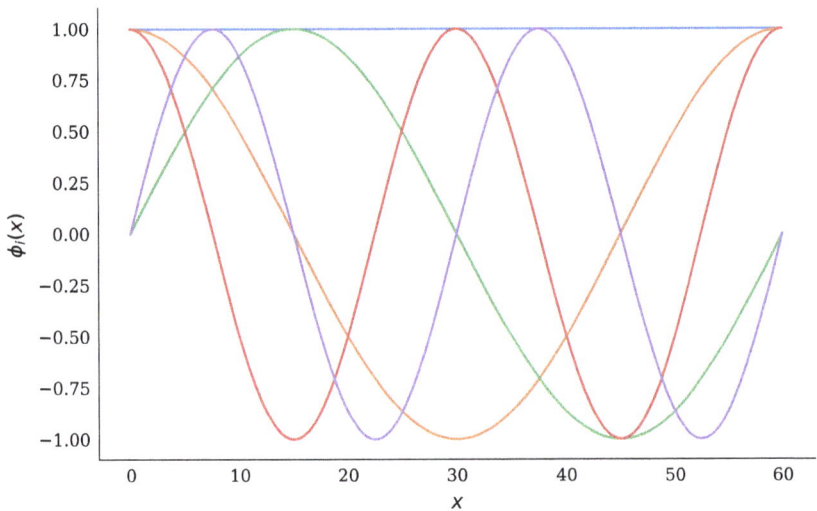

Figure 15.9. Fourier basis.

Figure 15.9 shows the result. We are going to express the acceleration of these basis functions.

We solve the least squares problem like this:

```
# Pick the parameters of the Fourier basis
L = 60.0
num_terms = 4
# Make the design matrix
Phi = get_fourier_design_matrix(X, L, num_terms)
# Solve least squares problem
w_LS = np.linalg.lstsq(Phi, Y, rcond=None)[0]
```

And we can make predictions like this:

```
# Make prediction at a dense set of points
xx = np.linspace(0, 60, 200)
Phi_xx = get_fourier_design_matrix(xx[:, None], L, num_terms)
Y_p = np.dot(Phi_xx, w_LS)
```

Figure 15.10 shows the result. As you can see, when we use just 4 basis functions, we overfit. Figure 15.11 shows the result when we use 20 (left) and 40 (right) basis functions. Notice that for 20 basis function, we obtain a good fit, but for 40 basis functions, we overfit.

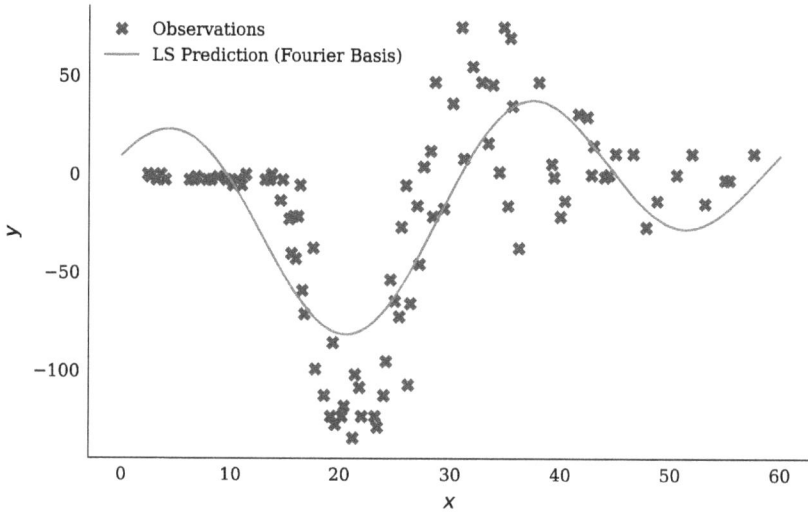

Figure 15.10. Fourier basis regression with 4 terms.

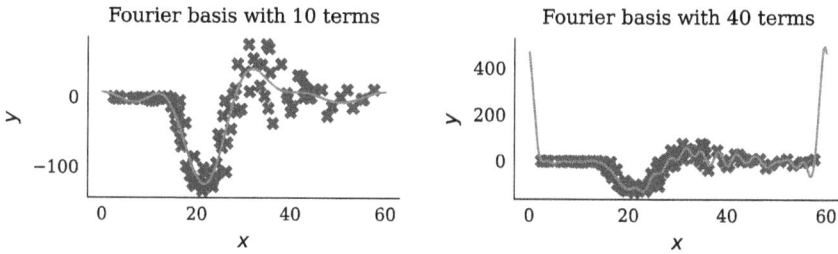

Figure 15.11. Fourier basis regression with 20 (left) and 40 (right) terms.

15.5 Example: Motorcycle Data with Radial Basis Functions

Let's now try out the radial basis functions. The mathematical form is:

$$\phi_i(x) = \exp\left\{ -\frac{(x - x_i^c)^2}{2\ell^2} \right\},$$

where x_i^c are points about each of the basis functions are centered and ℓ is the width of the basis function. We start with the code that evaluates the design matrix:

```
def get_rbf_design_matrix(x, x_centers, ell):
    """Radial basis functions design matrix.

    Arguments:
        x           -       the input points on which you want to
↪   evaluate the
                            design matrix
        x_center    -       the centers of the radial basis functions
        ell         -       the lengthscale of the radial basis
↪   function
    """
    # Make sure this is a 2D numpy array with only one column
    assert isinstance(x, np.ndarray), 'x is not a numpy array.'
    assert x.ndim == 2, 'You must make x a 2D array.'
    assert x.shape[1] == 1, 'x must be a column.'
    N = x.shape[0]
    cols = [np.ones((N, 1))]
    # Loop over columns and add the polynomial
    for i in range(x_centers.shape[0]):
        cols.append(np.exp(-(x - x_centers[i]) ** 2 / ell))
    return np.hstack(cols)
```

Notice that they are Gaussian functions centered at the points x_i^c. In this example, we are taking the x_i^c to be 10 points evenly spaced between 0 and 60. The width ℓ is 5. Figure 15.12 visualizes the basis functions. Figure 15.13 shows the result of the regression. In the hands-on activity, you will have to experiment with different widths and number of basis functions to find the best fit.

15.6   Measures of Predictive Accuracy: Colab

You cannot test how good your model is using the training dataset. Whatever the metric you use, the performance of your model on the training dataset will always be quite good. This is because the model is tuned to do well on the training data. The real question is how well your model does on a dataset it has never see. This brings us to the concept of a *test dataset*. How can you make a test dataset? Well,

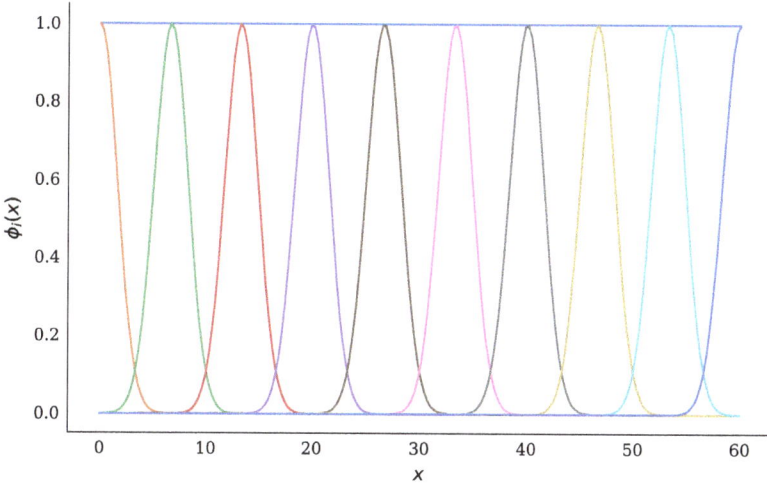

Figure 15.12. Radial basis functions.

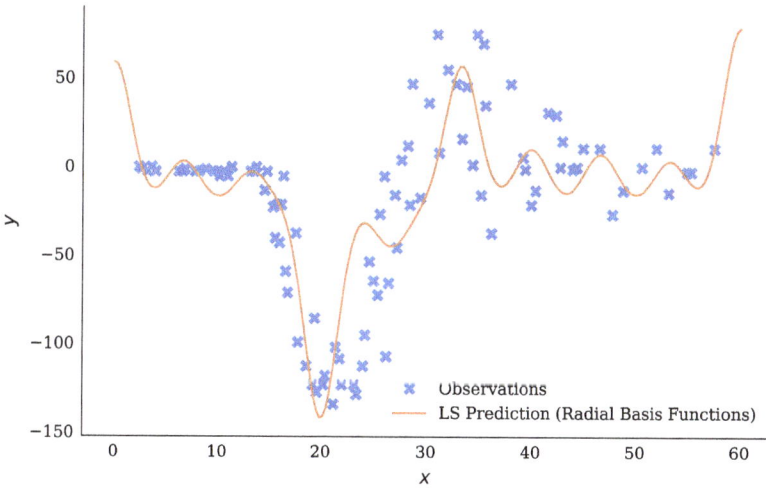

Figure 15.13. Regression with radial basis functions.

take whatever data you have and split them into training and test datasets. For example, you can randomly select 70% of your data and put it in your training set and 30% of the data and put it in your test set.

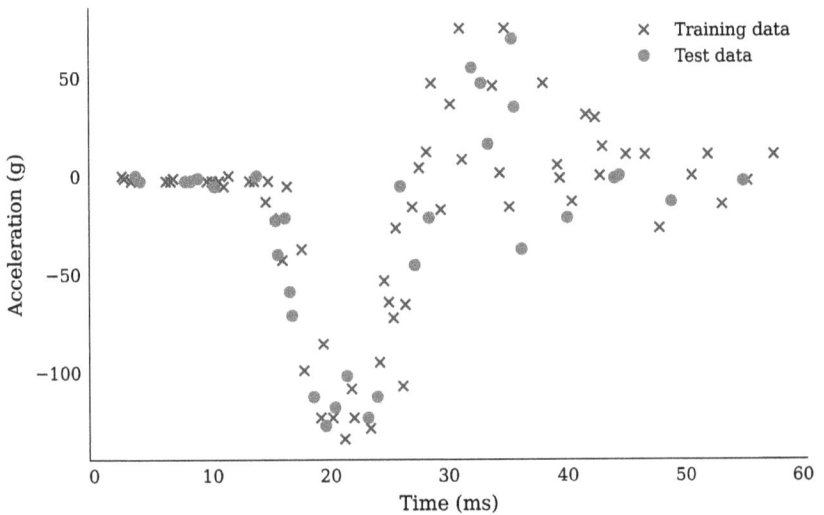

Figure 15.14.   Training and test datasets.

Let's do this for the motorcycle dataset. We are going to use the functionality of the Python library `scikit-learn` to split our dataset in to training and test. It is a lot easier than doing it manually.

```
from sklearn.model_selection import train_test_split
x_train, x_test, y_train, y_test = train_test_split(x, y,
↪   test_size=0.33)
```

Figure 15.14 visualizes the split.

The idea is to fit the model on the training set and then evaluate the performance on the test set. For example, Figure 15.15 shows the fit of the polynomial regression of degree 3 on the training and test sets. Throughout this section, we are going to demonstrate some of the most popular measures of predictive accuracy using this example.

15.6.1   *The predictions-vs-observations plot*

A very nice plot you can do is the predictions-vs-observations plot for the test data. Here is how:

```
Phi_poly_test = get_polynomial_design_matrix(x_test[:, None],
↪   degree)
y_test_predict = np.dot(Phi_poly_test, w_poly)
```

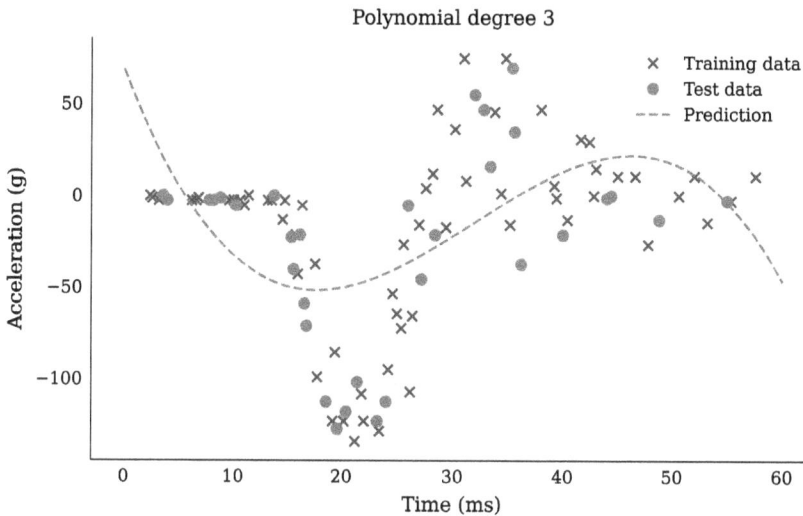

Figure 15.15. Polynomial regression of degree 3 on the training and test sets.

```
fig, ax = plt.subplots()
ax.set_title('Polynomial degree {0:d}'.format(degree))
ax.plot(y_test_predict, y_test, 'o')
yys = np.linspace(y_test.min(), y_test.max(), 100)
ax.plot(yys, yys, 'r-');
ax.set_xlabel('Predictions')
ax.set_ylabel('Observations')
```

Figure 15.16 shows the result. Notice that in this plot, I have included a line that goes through the origin at has a 45 degree angle with the x-axis. This line marks the perfect fit, i.e., a perfect agreement between the predictions and the observations. In other words, the closer the dots get to the line, the better the model.

15.6.2   *The mean squared error*

Sometimes you just want to characterize your model using a number. The mean squared error or (MSE) is such a scalar measure of the goodness of fit. The MSE is the *mean* of the *sum* of the *square* of the *prediction error* over your test data. If we assume that you have N_t test data points with inputs $x_{t,i}$ and outputs $y_{t,i}$, the MSE of your

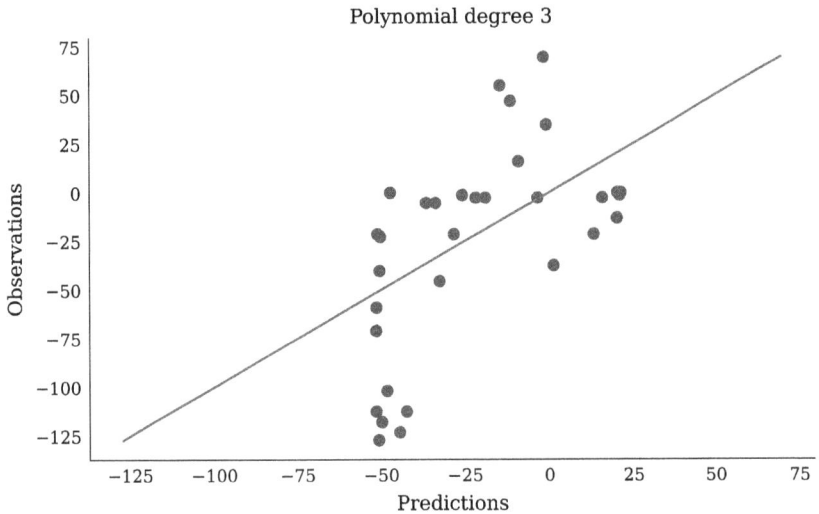

Figure 15.16. Predictions-vs-observations plot for the polynomial regression of degree 3.

model is given by the formula:

$$\text{MSE} = \frac{\sum_{i=1}^{N_t} \left[y_{t,i} - \mathbf{w}^T \phi(x_{t,i}) \right]^2}{N_t}.$$

In the polynomial model we fitted above, it is:

```
MSE_poly = np.mean((y_test_predict - y_test) ** 2)
print('MSE_poly = {0:1.2f}'.format(MSE_poly))
```

The output is:

```
MSE_poly = 1765.92
```

Now, it is obvious that smaller MSE means better model. But it is a bit hard to understand the absolute meaning of the MSE. So, sometimes we look at the relative MSE or (RMSE).[1] RMSE is the MSE of your model divided by the MSE of the simplest possible model you could make. Okay. What is the simplest possible model you could make? It is a constant model, i.e., A model that predicts

---

[1] Note that sometimes the acronym RMSE is used for the root mean square error which is just the square root of the MSE.

the same value for the output y no matter what the input x is. So, mathematically the model is:

$$y_{\text{simplest}} = c,$$

for some constant c. What is the best such constant? Well, you can fit it by minimizing the sum of square errors for this simple model. If you do it, you will get, very intuitively, that the constant should be the empirical average of your training data, i.e.,

$$y_{\text{simplest}} = \hat{\mu} = \frac{1}{N} \sum_{i=1}^{N} y_i.$$

So, the MSE of this simple model is just:

$$\text{MSE}_{\text{simplest}} = \frac{\sum_{i=1}^{N_t} (y_{t,i} - \hat{\mu})^2}{N_t}.$$

And we define RMSE to be:

$$\text{RMSE} = \frac{\text{MSE}}{\text{MSE}_{\text{simplest}}} = \frac{\sum_{i=1}^{N_t} \left[y_{t,i} - \mathbf{w}^T \phi(x_{t,i})\right]^2}{\sum_{i=1}^{N_t} (y_{t,i} - \hat{\mu})^2}.$$

The RMSE has an intuitive meaning. If it is smaller than one, then this means that your model is doing something better than the simplest possible model. And the smaller it is, the better you do compared to the simplest model. If the RMSE is greater than one, then your model is really, really bad.

Let's find the RMSE for our case:

```
mu = y_train.mean()
MSE_simplest = np.mean((y_test - mu) ** 2)
print('MSE_simplest = {0:1.2f}'.format(MSE_simplest))
RMSE_poly = MSE_poly / MSE_simplest
print('RMSE_poly = {0:1.2f}'.format(RMSE_poly))
```

The output is:

```
MSE_simplest = 2728.85
RMSE_poly = 0.65
```

15.6.3 *Coefficient of determination R^2*

The coefficient of determination is defined by:

$$R^2 = 1 - \text{RMSE} = 1 - \frac{\sum_{i=1}^{N_t} \left[y_{t,i} - \mathbf{w}^T \phi(x_{t,i}) \right]^2}{\sum_{i=1}^{N_t} (y_{t,i} - \hat{\mu})^2}.$$

This measure is telling you how much of the percentage of the variance of the test data is explained by your model. So, here you want to get as close to one as possible.

Let's see what we get in our case:

```
R2_poly = 1 - RMSE_poly
print('R2 = {0:1.2f}'.format(R2_poly))
```

The output is:

```
R2 = 0.35
```

So, in this example our model explains about 35% of the variance of the test data? What is the rest? Well, the rest explained by the measurement noise. This concept is more advanced than my intention for this book.

15.6.4 *Example: Repeat the fitting using Fourier features*

Let's repeat all these diagnostics for the Fourier basis with 20 terms. The left panel of Figure 15.17 shows the fit of the Fourier basis with 20 terms on the training and test sets. The right panel of Figure 15.17 shows the predictions-vs-observations plot for the test set.

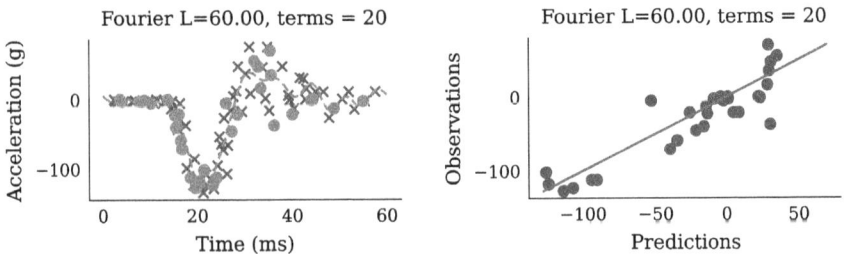

Figure 15.17. Fourier basis with 20 terms on the training and test sets (left) and predictions-vs-observations plot for the test set (right).

The statistics are:

```
MSE_fourier = 523.75
RMSE_fourier = 0.19
R2 = 0.81
```

Notice that this model does much better than the polynomial model by all measures. In particular, R2 is telling us that 81% of the variance of the test data is explained by our model.

15.7 Cross Validation for Selecting the Number of Basis Functions: Colab

In the previous section, we saw that the Fourier basis gave us a much better model for the motorcycle dataset. But, how do we know how many basis functions should we choose? Can we automate the process of selecting the number of basis functions?

To answer this question, we are going to use a technique known as *cross-validation*. The idea is simple. Pick the number of basis functions that give you on average the smallest validation error. By *validation error*, we mean the error on a subset of data not used in training. By *average*, we mean averaging over the possible ways in which you can split your dataset into training and validation.

Note that the validation dataset should be completely distinct from your test dataset. So, the process goes like this:

- You split the original dataset into training and test subsets. You set the test aside for final diagnostics.
- You repeatedly split the training subset into two subsets: one for training and one for validation.
- You fit the model on the training subset and evaluate the MSE on the validation subset.
- You average the MSE over all possible splits.
- You pick the model with the smallest average MSE.

Let me demonstrate directly on the motorcycle data. For convenience, we are going to use the `RepeatedKFold` class from the `sklearn.model_selection` module. This class implements the cross-validation procedure we described above. We are going to use 10-fold cross-validation with 10 repeats. The 10-fold means that we will split the training data into 10 folds and we will use 9 of them for training and 1 for validation. The 10 repeats mean that we will repeat the process 10 times and average the MSE. The code is as follows:

```
from sklearn.model_selection import RepeatedKFold
# This will hold the average MSE for each possible number of basis
↪  functions
MSE = []
fourier_L = 60.0
# This is the cross-validation object
rkf = RepeatedKFold(n_splits=10, n_repeats=10)
for fourier_terms in range(1, 20):
    # This will hold the sum of the MSEs for this number of basis
    ↪  functions
    mse_sum = 0.0
    for train_index, valid_index in rkf.split(x_train):
        # Extract training/validation features
        x_train_train, x_train_valid = x_train[train_index],
        ↪  x_train[valid_index]
        # Extract training/validation targets
        y_train_train, y_train_valid = y_train[train_index],
        ↪  y_train[valid_index]
        # Get the design matrix for the training set
        Phi_fourier_train =
        ↪  get_fourier_design_matrix(x_train_train[:, None],
        ↪  fourier_L, fourier_terms)
        # Fit the model on the training set
        w_fourier, _, _, _ = np.linalg.lstsq(Phi_fourier_train,
        ↪  y_train_train, rcond=None)
        # Get the design matrix for the validation set
        Phi_fourier_valid =
        ↪  get_fourier_design_matrix(x_train_valid[:, None],
        ↪  fourier_L, fourier_terms)
        # Make predictions on the validation set
        y_valid_predict = np.dot(Phi_fourier_valid, w_fourier)
        # Compute the MSE on the validation set
        MSE_fourier = np.mean((y_valid_predict - y_train_valid) **
        ↪  2)
        # Add the MSE to the sum
```

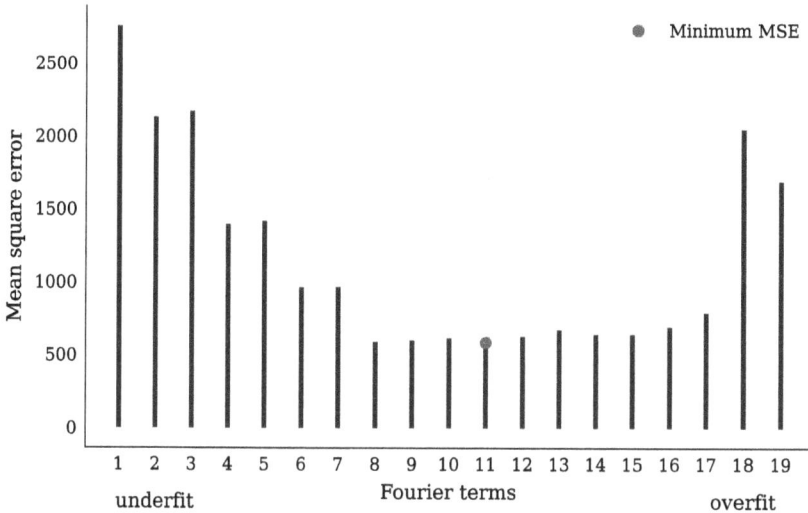

Figure 15.18. Cross-validation for selecting the number of basis functions.

```
mse_sum += MSE_fourier
# Average the MSE over all the folds and repeats
MSE.append(mse_sum / rkf.get_n_splits())
```

Figure 15.18 shows the result. The minimum MSE is at 11 basis functions. Notice that we can clearly see the underfit and overfit regions for too few and too many basis functions.

15.8 Maximum Likelihood Interpretation of Least Squares

I now show you how to derive least squares from the maximum likelihood principle. Recall that the maximum likelihood principle states that you should pick the model parameters that maximize the probability of the data conditioned on the parameters.

Just like before assume that we have N observations of inputs $\mathbf{x}_{1:N}$ and outputs $\mathbf{y}_{1:N}$. We model the map between inputs and outputs using a generalized linear model with M basis functions:

$$y(\mathbf{x}; \mathbf{w}) = \sum_{j=1}^{M} w_j \phi_j(\mathbf{x}) = \mathbf{w}^T \phi(\mathbf{x})$$

Now here is the difference with what we did before. Instead of directly picking a loss function to minimize, we come up with a probabilistic description of the measurement process. In particular, we model the measurement process using a likelihood function:

$$\mathbf{y}_{1:N}|\mathbf{x}_{1:N}, \mathbf{w} \sim p(\mathbf{y}_{1:N}|\mathbf{x}_{1:N}, \mathbf{w}).$$

What is the interpretation of the likelihood function? Well, $p(\mathbf{y}_{1:N}|\mathbf{x}_{1:N}, \mathbf{w})$ tells us how plausible is it to observe $\mathbf{y}_{1:N}$ at inputs $\mathbf{x}_{1:N}$, if we know that the model parameters are $\mathbf{w}$.

The most common choice for the likelihood of a single measurement is to pick it to be Normal. This corresponds to the belief that our measurement is around the model prediction $\mathbf{w}^{\mathbf{T}}\phi(\mathbf{x})$ but it is contaminated with Gaussian noise of variance σ^2. Mathematically, we have:

$$p(y_i|\mathbf{x}_i, \mathbf{w}, \sigma) = N\left(y_i|y(\mathbf{x}_i; \mathbf{w}), \sigma^2\right)$$
$$= N\left(y_i|\mathbf{w}^T\phi(\mathbf{x}_i), \sigma^2\right),$$

where σ^2 models the *variance of the measurement noise*. Note that here I used the notation $N(y|\mu, \sigma^2)$ to denote the PDF of a Normal with mean μ and variance σ^2, i.e.,

$$N(y|\mu, \sigma^2) := (2\pi\sigma^2)^{-\frac{1}{2}} \exp\left\{-\frac{(y-\mu)^2}{2\sigma^2}\right\}.$$

Since, in almost all the cases we encounter, the measurements are independent conditioned on the model, then the likelihood of the data factorizes as follows:

$$p(\mathbf{y}_{1:N}|\mathbf{x}_{1:N}, \mathbf{w}) = \prod_{i=1}^{N} p(y_i|\mathbf{x}_i, \mathbf{w})$$

$$= \prod_{i=1}^{N} N\left(y_i|\mathbf{w}^{\mathbf{T}}\phi(\mathbf{x_i}), \sigma^2\right)$$

$$= \prod_{i=1}^{N} (2\pi\sigma^2)^{-\frac{1}{2}} \exp\left\{-\frac{\left[y_i - \mathbf{w}^T\phi(\mathbf{x}_i)\right]^2}{2\sigma^2}\right\}$$

$$= (2\pi\sigma^2)^{-\frac{N}{2}} \exp\left\{ -\sum_{i=1}^{N} \frac{\left[y_i - \mathbf{w}^T \phi(\mathbf{x}_i) \right]^2}{2\sigma^2} \right\}$$

$$= (2\pi\sigma^2)^{-\frac{N}{2}} \exp\left\{ -\frac{1}{2\sigma^2} \sum_{i=1}^{N} \left[y_i - \mathbf{w}^T \phi(\mathbf{x}_i) \right]^2 \right\}$$

$$= (2\pi\sigma^2)^{-\frac{N}{2}} \exp\left\{ -\frac{1}{2\sigma^2} \| \mathbf{y}_{1:N} - \mathbf{\Phi}\mathbf{w} \|^2 \right\},$$

where $\mathbf{\Phi}$ is the $N \times M$ design matrix.

Now we are ready to apply the maximum likelihood function to find all the parameters. This includes both the weight vector $\mathbf{w}$ and the measurement variance σ^2. We need to solve this:

$$\max_{\mathbf{w},\sigma^2} \log p(\mathbf{y}_{1:N}|\mathbf{x}_{1:N}, \mathbf{w})$$

$$= \max_{\mathbf{w},\sigma^2} \left\{ -\frac{N}{2} \log(2\pi) - \frac{N}{2} \log \sigma^2 - \frac{1}{2\sigma^2} \| \mathbf{y}_{1:N} - \mathbf{\Phi}\mathbf{w} \|^2 \right\}$$

Notice that the rightmost part is actually the negative of the sum of square errors. So, by maximizing the likelihood with respect to $\mathbf{w}$, we are actually minimizing the sum of square errors. This means that the maximum likelihood weights and the least square weights are exactly the same! We do not even have to do anything further. The weights should satisfy this linear system:

$$\mathbf{\Phi}^T \mathbf{\Phi}\mathbf{w} = \mathbf{\Phi}^T \mathbf{y}_{1:N}.$$

This is nice! The probabilistic interpretation above gives the same solution as least squares! But there is more. Notice that it can also give us an estimate for the measurement noise variance σ^2. All you have to do is maximize likelihood with respect to σ^2. If we take the derivative of the log-likelihood with respect to σ^2, set it equal to zero and solve for σ^2 you get:

$$\sigma^2 = \frac{\| \mathbf{\Phi}\mathbf{w} - \mathbf{y}_{1:N} \|^2}{N}.$$

Finally, you can incorporate this measurement uncertainty when you are making predictions. This is done through the *point predictive distribution*, which is Normal in our case:

$$p(y|\mathbf{x}, \mathbf{w}, \sigma^2) = \mathcal{N}\left(y|\mathbf{w}^T\phi(\mathbf{x}), \sigma^2\right).$$

In other words, your prediction about the measured output y is that it will be Normally distributed around your model prediction with a variance σ^2. You can use this to find a 95% credible interval. Let's demonstrate with an example.

15.8.1 *Example: Maximum likelihood interpretation of least squares: Colab*

We are going to use the quadratic synthetic data from Section 15.2.1. We fit a quadratic model, just like before:

```
# The degree of the polynomial
degree = 2
# The design matrix
Phi = get_polynomial_design_matrix(x[:, None], degree)
# Solve the least squares problem
w, sum_res, _, _ = np.linalg.lstsq(Phi, y, rcond=None)
```

Notice now that I am picking up the `sum_res` output of the `np.linalg.lstsq` function which I was ignoring before. This is the sum of square errors, i.e.,

$$\sum_{i=1}^{N}\left[y_i - \sum_{j=1}^{M} w_j\phi_j(x_i)\right]^2 = \|\mathbf{y}_{1:N} - \mathbf{\Phi}\mathbf{w}\|_2^2.$$

We can estimate the variance and the standard deviation of the measurement noise by:

```
sigma2_MLE = sum_res[0] / num_obs
sigma_MLE = np.sqrt(sigma2_MLE)
print(f'True sigma = {sigma_true:1.4f}')
print(f'MLE sigma = {sigma_MLE:1.4f}')
```

The output is:

```
True sigma = 0.1000
MLE sigma = 0.1000
```

We can make predictions with uncertainty as follows:

```
fig, ax = plt.subplots()
# Some points on which to evaluate the regression function
xx = np.linspace(-1, 1, 100)
# The true connection between x and y
yy_true = w0_true + w1_true * xx + w2_true * xx ** 2
# The mean prediction of the model we just fitted
Phi_xx = get_polynomial_design_matrix(xx[:, None], degree)
yy = np.dot(Phi_xx, w)
# Lower bound for 95% credible interval
sigma_MLE = np.sqrt(sigma2_MLE)
yy_l = yy - 2.0 * sigma_MLE
# Upper bound for 95% credible interval
yy_u = yy + 2.0 * sigma_MLE
# plot mean prediction
ax.plot(xx, yy, '--', label='Mean prediction')
# plot shaded area for 95% credible interval
ax.fill_between(xx, yy_l, yy_u, alpha=0.25, label='95% credible
↪   interval')
# plot the data again
```

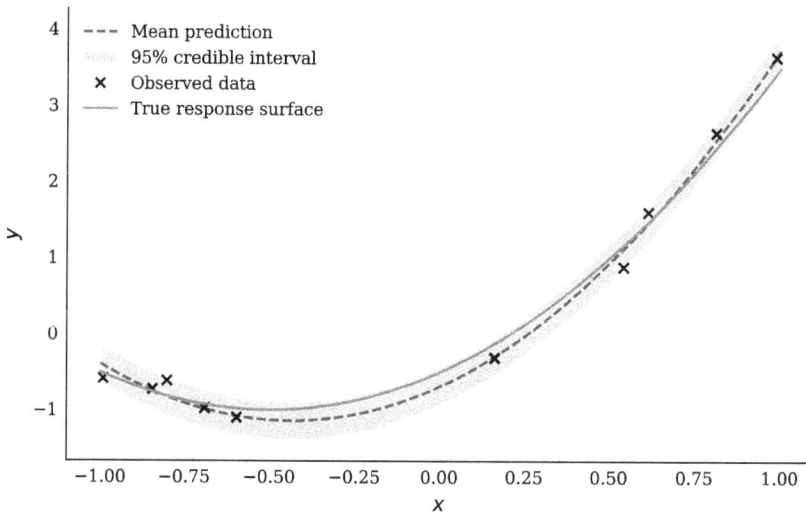

Figure 15.19. Maximum likelihood interpretation of least squares.

```
ax.plot(x, y, 'kx', label='Observed data')
# overlay the true
ax.plot(xx, yy_true, label='True response surface')
ax.set_xlabel('$x$')
ax.set_ylabel('$y$')
plt.legend(loc='best')
```

The result is shown in Figure 15.19.

Exercises

1. **Optimizing the performance of a compressor.** In this problem, you are going to need the `compressor_data.xlsx` dataset from the books repository. The dataset was kindly provided to us by Professor Davide Ziviani. Once you have downloaded the file, you can load it like this:

```
import pandas as pd
data = pd.read_excel('compressor_data.xlsx')
```

The data are part of an experimental study of a variable speed reciprocating compressor. The experimentalists varied two temperatures T_e and T_c (both in degrees Celsius) and they measured various other quantities. Our goal is to learn the map between T_e and T_c and measured Capacity and Power (both in Watts). First, let's see how you can extract only the relevant data to perform regression between T_e and T_c and measured Capacity.

```
X = data[['T_e','T_c']].values
y = data['Capacity'].values
```

(a) Fit the following multivariate polynomial model to both the Capacity:

$$y = w_1 + w_2 T_e + w_3 T_c + w_4 T_e T_c + w_5 T_e^2 + w_6 T_c^2$$
$$+ w_7 T_e^2 T_c + w_8 T_e T_c^2 + w_9 T_e^3 + w_{10} T_c^3.$$

You may use `sklearn.preprocessing.PolynomialFeatures` to construct the design matrix of your polynomial features. Do not program the design matrix by hand. You should split your data into training and test and use various diagnostics to make sure that your models make sense. This code can get you started:

```
# For polynomial features
from sklearn.preprocessing import PolynomialFeatures
# For splitting the data
from sklearn.model_selection import train_test_split
# Split the data
X_train, X_test, y_train, y_test = train_test_split(X, y,
↪    test_size=0.33)
# Here is how to make Polynomials features
poly = PolynomialFeatures(degree=3)
# Design matrix for train data
Phi_train = poly.fit_transform(X_train)
```

Find the mean squared error, the coefficient of determination R^2 of the test data. Make the predictions-vs-observations plot for the test data.
(b) Repeat the process for the Power.
(c) Use 10-fold cross-validation with 10 repeats to select the maximum polynomial degree.
(d) Use the maximum likelihood interpretation of least squares to estimate the variance of the measurement noise.

2. **Data standardization.** Out of the box regression methods often make the assumption that the data are centered around zero and have unit variance. This is not always the case, e.g., the motorcycle dataset. In this problem, you will learn how to *standardize* the data. This is something that you should almost always do before fitting a model. The idea is simple. You compute the empirical mean $\hat{\mu}$ and standard deviation $\hat{\sigma}$ of the data. Then you take each observation, you subtract the mean and divide by the standard deviation. Like this:

$$x_i^s = \frac{x_i - \hat{\mu}}{\hat{\sigma}}.$$

You do this for each dimension of the input and also for the output.

(a) Load and standardize the motorcycle dataset. Do it manually using standard numpy functions. Visualize your standardized data using a scatter plot.

(b) Repeat the standardization process using the `sklearn.preprocessing.StandardScaler`.

(c) Split your data into training and test.

(d) Fit a polynomial model of degree 5 to the standardized training data. Call this model `model_S`.

(e) Write a function takes an original input, standardizes it, makes a prediction with `model_S`, and then unstandardizes the prediction. Call this function `predict_unstandardized`. Use the function to compute the MSE on the unstandardized test data.

(f) Fit a polynomial model of degree 5 to the unstandardized training data and compute the MSE on the unstandardized test data.

(g) Compare the MSE of the two models. Is there a difference?

Chapter 16

Classification via Logistic Regression

In this chapter, I introduce the classification problem and solve it using logistic regression. I show how to formulate the parameter estimation problem via maximum likelihood. I demonstrate that maximizing the likelihood is the same as minimizing the, so-called, entropy loss. I introduce diagnostics for validating a logistic regression model.

16.1 Logistic Regression

Imagine that you have a bunch of observations consisting of inputs $\mathbf{x}_{1:N} = (\mathbf{x}_1, \ldots, \mathbf{x}_N)$ and the corresponding outputs $y_{1:N} = (y_1, \ldots, y_N)$. When y_i is a discrete label, we have a *classification* problem. In particular, if the labels are two, say 0 or 1, then we say that we have a *binary classification* problem.

The logistic regression model is one of the simplest ways to solve the binary classification problem. It goes as follows. You model the probability that $y = 1$ conditioned on having $\mathbf{x}$ by:

$$p(y = 1|\mathbf{x}, \mathbf{w}) = \text{sigm}\left(\sum_{j=1}^{M} w_j \phi_j(\mathbf{x})\right) = \text{sigm}\left(\mathbf{w}^T \boldsymbol{\phi}(\mathbf{x})\right),$$

where sigm is the sigmoid function, the $\phi_j(\mathbf{x})$ are m basis functions/features,

$$\boldsymbol{\phi}(\mathbf{x}) = (\phi_1(\mathbf{x}), \ldots, \phi_M(\mathbf{x}))$$

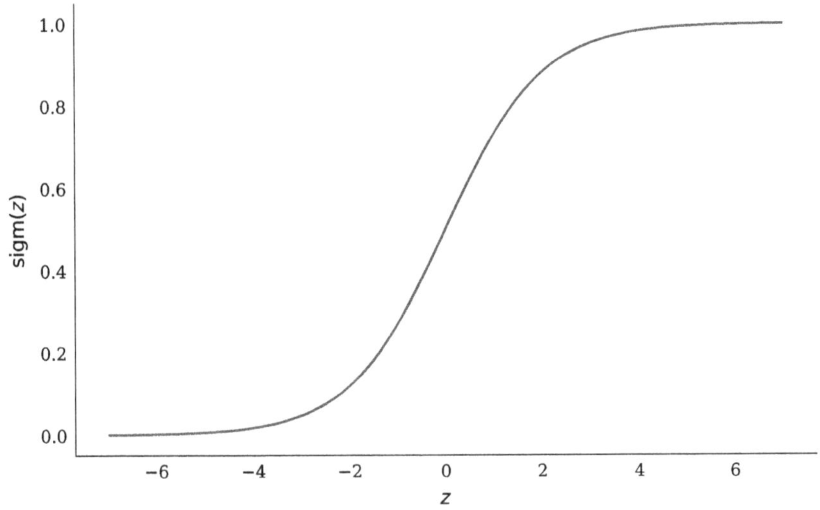

Figure 16.1. The sigmoid function takes any number and maps it to the unit interval $[0, 1]$.

and the w_j's are m weights that we need to learn from the data. The sigmoid function is defined by:

$$\text{sigm}(z) = \frac{1}{1 + e^{-z}},$$

and all it does is it takes a real number and maps it to $[0, 1]$ so that it can represent a probability, see Figure 16.1. In other words, logistic regression is just a generalized linear model passed through the sigmoid function so that it is mapped to $[0, 1]$.

If you need the probability of $y = 0$, it is given by the obvious rule:

$$p(y = 0 | \mathbf{x}, \mathbf{w}) = 1 - p(y = 1 | \mathbf{x}, \mathbf{w}) = 1 - \text{sigm}\left(\mathbf{w}^T \phi(\mathbf{x})\right)$$

You can represent the probability of an arbitrary label y conditioned on $\mathbf{x}$ using this simple trick:

$$p(y | \mathbf{x}, \mathbf{w}) = \left[\text{sigm}\left(\mathbf{w}^T \phi(\mathbf{x})\right)\right]^y \left[1 - \text{sigm}\left(\mathbf{w}^T \phi(\mathbf{x})\right)\right]^{1-y}.$$

Notice that when $y = 1$, the exponent of the second term becomes zero and thus the term becomes one. Similarly, when $y = 0$, the exponent of the first term becomes zero and thus the term becomes one. This gives the right probability for each case. The likelihood of all the observed data is:

$$p(y_{1:N}|\mathbf{x}_{1:n}, \mathbf{w}) = \prod_{i=1}^{N} p(y_i|\mathbf{x}_i, \mathbf{w})$$

$$= \prod_{i=1}^{N} \left[\text{sigm}\left(\mathbf{w}^T\phi(\mathbf{x}_i)\right)\right]^{y_i} \left[1 - \text{sigm}\left(\mathbf{w}^T\phi(\mathbf{x}_i)\right)\right]^{1-y_i}.$$

We can now find the best weight vector $\mathbf{w}$ using the maximum likelihood principle. We need to solve the optimization problem:

$$\max_{\mathbf{w}} \log p(y_{1:N}|\mathbf{x}_{1:n}, \mathbf{w}) = \max_{\mathbf{w}} \sum_{i=1}^{N} \left\{y_i \, \text{sigm}\left(\mathbf{w}^T\phi(\mathbf{x}_i)\right)\right.$$

$$\left. + (1 - y_i)\left[1 - \text{sigm}\left(\mathbf{w}^T\phi(\mathbf{x}_i)\right)\right]\right\}.$$

Notice that the following maximization problem is equivalent to minimizing this loss function:

$$L(\mathbf{w}) = -\sum_{i=1}^{N} \left\{y_i \, \text{sigm}\left(\mathbf{w}^T\phi(\mathbf{x}_i)\right) + (1 - y_i)\left[1 - \text{sigm}\left(\mathbf{w}^T\phi(\mathbf{x}_i)\right)\right]\right\}.$$

This is known as the *cross-entropy loss function* and you are very likely to encounter it if you dive deeper into modern data science and machine learning. For example, we use the same loss function to train state-of-the-art deep neural networks that classify images. You now know that it does not come out of the blue. It comes from the maximum likelihood principle.[1]

---

[1]It is not possible to analytically minimize the cross-entropy loss function. The optimization is usually done with a numerical optimization technique. For small datasets, we can use a gradient-based optimization (e.g., the conjugate gradient method, or the Newton–Raphson method). The details of this optimization algorithm are beyond the scope of the present course. For large datasets, this is typically the case for deep learning, we use variants of stochastic gradient descent.

330 Introduction to Data Science for Engineering Students

16.2 Example: Logistic Regression with One Variable (High Melting Explosives): Colab

High Melting Explosives (HMX) have applications as detonators of nuclear weapons and as solid rocket propellants. We will use logistic regression to build the probability that a specific HMX block explodes when dropped from a given height. To this end, we will use data the dataset `hmx_data.csv`, see Casey (2020) and Smith (1987).

```
import pandas as pd
df = pd.read_csv('hmx_data.csv')
df.head()
```

Some data:

```
   Height Result
0   40.5   E
...
47  25.5   N
...
```

Each row of the data is a different experiment. There are two columns:

- First column is height: From what height (in cm) was the specimen dropped from.
- Second column is result: Did the specimen explode (E) or not (N)?

Let's encode the labels as 1 and 0 instead of E and N. Let's do this below:

```
x = data['Height'].values
label_coding = {'E': 1, 'N': 0}
y = np.array([label_coding[r] for r in data['Result']])
data['y'] = y
```

This just added a new column, called `y`, to the dataframe.

We can now train a logistic regression model with just a linear feature using `scikit-learn`:

```
from sklearn.linear_model import LogisticRegression
X = x[:, None]
model = LogisticRegression().fit(X, y)
```

Here is how you can make predictions at some arbitrary heights:

```
X_predict = np.array([10.0, 20.0, 30.0, 40.0, 50.0])[:, None]
predictions = model.predict_proba(X_predict)
predictions.round(2)
```

Output:

```
array([[1.  , 0.  ],
       [0.99, 0.01],
       [0.58, 0.42],
       [0.02, 0.98],
       [0.  , 1.  ]])
```

The first column is the probability of not exploding, the second column is the probability of exploding.

If you wanted, you could ask for the class of maximum probability for each prediction input:

```
model.predict(X_predict)
```

Output:

```
array([0, 0, 0, 1, 1])
```

To visualize the predictions of the model as a function of the height, we can do this:

```
fig, ax = plt.subplots()
xx = np.linspace(20.0, 45.0, 100)
XX = xx[:, None]
predictions_xx = model.predict_proba(XX)
ax.plot(xx, predictions_xx[:, 0], label='Probability of N')
ax.plot(xx, predictions_xx[:, 1], label='Probability of E')
ax.set_xlabel('$x$ (cm)')
ax.set_ylabel('Probability')
plt.legend(loc='best')
```

Figure 16.2 shows the predictions of the model as a function of the height.

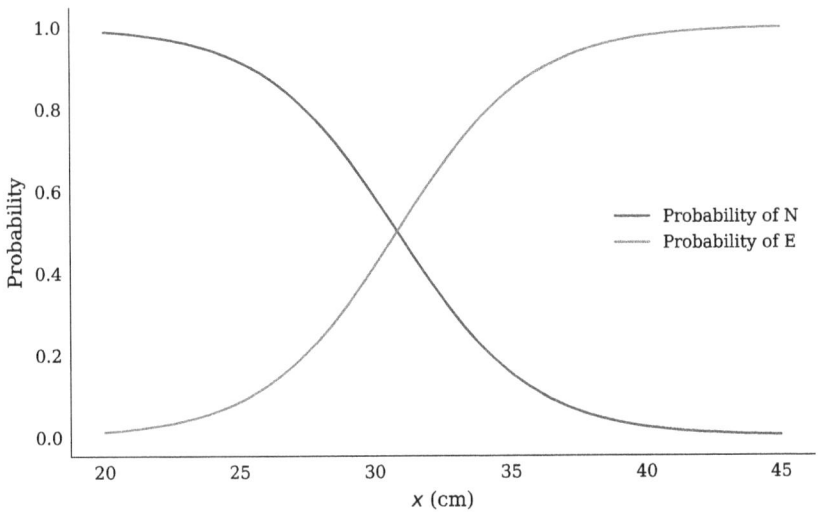

Figure 16.2. Predictions of the logistic regression model as a function of the height.

16.3   Logistic Regression with Many Features: Colab

Let's repeat what we did for the HMX example. Instead of using a linear model inside the sigmoid, we will use a quadratic model. That is, the probability of an explosion will be:

$$p(y = 1|x, \mathbf{w}) = \text{sigm} \left(w_0 + w_1 x + w_2 x^2 \right).$$

This is how easy it is to do this in `scikit-learn`:

```
from sklearn.preprocessing import PolynomialFeatures
from sklearn.linear_model import LogisticRegression
poly = PolynomialFeatures(2)
Phi = poly.fit_transform(x[:, None])
model = LogisticRegression().fit(Phi, y)
```

The coefficient for the quadratic term is almost zero, so there the model is almost exactly as the one we trained with a linear feature.

16.4 Diagnostics for Classification: Colab

Let's repeat what we did for the HMX example, but after splitting the dataset into training and test subsets. We will be making predictions on the test subset.

```
# Split the data into training and test subsets
from sklearn.model_selection import train_test_split
x_train, x_test, y_train, y_test = train_test_split(x, y)
# Train the model
from sklearn.preprocessing import PolynomialFeatures
from sklearn.linear_model import LogisticRegression
poly = PolynomialFeatures(2)
Phi_train = poly.fit_transform(x_train[:, None])
model = LogisticRegression().fit(Phi_train, y_train)
# Make predictions on the test subset
Phi_test = poly.fit_transform(x_test[:, None])
predictions = model.predict_proba(Phi_test)
```

As we have already mentioned, the predictions are probabilistic. One of the easiest visualizations you can do is the following. In the same plot, show put a point either at 0 or 1 depending on what the actual prediction is and then draw a bar for the probability that $y = 1$. Here it is:

```
fig, ax = plt.subplots()
ax.bar(np.arange(y_test.shape[0]), predictions[:, 1], color='g',
↪   alpha=0.5, label='$p(y=1|x,w)$')
ax.plot(y_test, 'o', label='Observed test point')
ax.set_xlabel('Test point id')
plt.legend(loc='best')
```

Figure 16.3 shows the result of this code. In this plot, the closer the bar is to the observation, the better the prediction.

Now, sometimes you may want to make pointwise predictions. That is, you look at the probability reported by the model and you

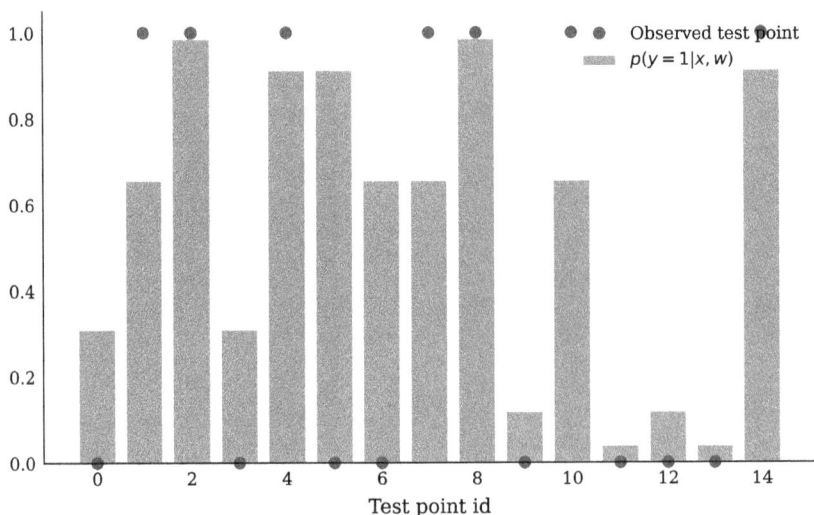

Figure 16.3. Predictions of the logistic regression model as a function of the height.

have to pick either 0 or 1.[2] Here, we are going to do something very simple. We will just pick the label that has the maximum probability. Here is how:

```
y_pred = np.argmax(predictions, axis=1)
```

Let's repeat the figure above, but now add this point prediction as a red cross, see Figure 16.4. The model is pretty good. Most of the times we make the correct prediction.

You can summarize how good the predictions are using the accuracy score. The accuracy score is:

$$\text{accuracy score} = \frac{\text{number of correct predictions}}{\text{total number of test samples}}.$$

Here is how we can calculate it:

```
from sklearn.metrics import accuracy_score
acc = accuracy_score(y_test, y_pred)
print(f'HMX Accuracy = {acc * 100:1.2f} %')
```

---

[2]There is a whole theory of how to do this properly: the *theory of decision-making* (Raiffa and Schlaifer, 1968).

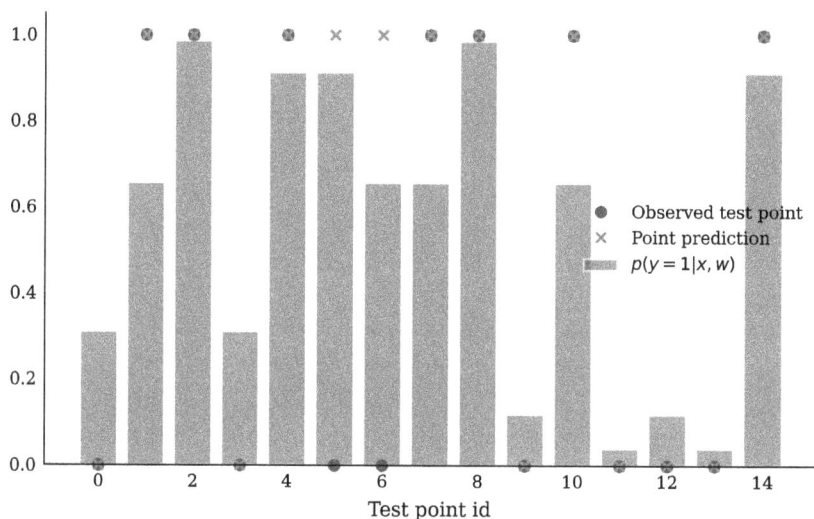

Figure 16.4. Predictions of the logistic regression model as a function of the height, including the point predictions.

Output:

```
HMX Accuracy = 86.67 %
```

Note that you cannot always have a model that is 100% accurate for many reasons. It may be the case that you don't have enough data to train a good model. But there is also a more fundamental reason. The situation you are studying may be stochastic, in the sense that the label is not a deterministic function of the input. In our example, this means that there are some heights that is really impossible to predict whether the block will explode or not.

Furthermore, the accuracy score can be very misleading. The problem is how well-balanced your dataset is. As an extreme example, imagine that we had a dataset in which explosions are very rare, say only 1%. Then a stupid model that predicts that there will never be an explosion would have 99% accuracy. This is clearly ridiculous. The situation is remedied by the concept of *balanced accuracy* (Brodersen *et al.*, 2010).

Finally, a nice way to visualize the predictions of your model is to use the *confusion matrix*. It is self-evident what it is when I plot it for you. Here you go:

```
from sklearn.metrics import confusion_matrix, ConfusionMatrixDisplay
fig, ax = plt.subplots()
cm = confusion_matrix(y_test, y_pred)
disp = ConfusionMatrixDisplay(confusion_matrix=cm)
disp.plot(ax=ax)
```

Figure 16.5 shows the confusion matrix. Here, the rows are the actual labels and the columns are the predicted labels. The diagonal elements are the number of correct predictions. The off-diagonal elements are the number of incorrect predictions. The more counts you have on the diagonal, the better the model. But the off-diagonal elements are also important. They tell you what is your model getting wrong. And, in this particular case, getting things wrong has asymetrically different costs. If you predict that there will be an explosion and there is not explosion, that's not a big deal. But if you predict that there will not be an explosion and there is one, that's a big deal.

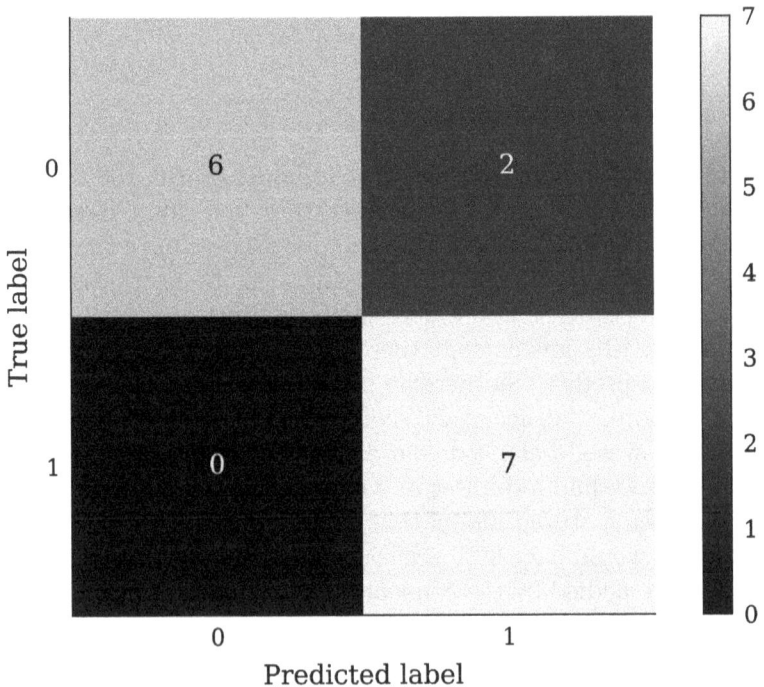

Figure 16.5. Confusion matrix for the HMX dataset.

So, you may want to penalize the false negatives more than the false positives.

Exercises

1. **Explaining the Challenger disaster.** On January 28, 1986, the Space Shuttle Challenger disintegrated after 73 seconds from launch. The failure can be traced on the rubber O-rings which were used to seal the joints of the solid rocket boosters (required to force the hot, high-pressure gases generated by the burning solid propellant through the nozzles thus producing thrust).

   It turns out that the performance of the O-ring material was particularly sensitive on the external temperature during launch. The `challenger_data.csv` dataset contains records of different experiments with O-rings recorded at various times between 1981 and 1986. Download the data the usual way. Even though this is a csv file, you should load it with pandas because it contains some special characters.

   The first column of the dataset is the date of the record. The second column is the external temperature of that day in degrees Fahrenheit. The third column labeled `Damage Incident` is has a binary coding (0 corresponds to "no damage", and 1 corresponds to "damage"). The very last row is the day of the Challenger accident.

   You are going to use the first 23 rows to solve a binary classification problem that will give you the probability of an accident conditioned on the observed external temperature in degrees Fahrenheit. Before you proceed to the analysis of the data, let's clean the data up.

   (a) Drop all the bad records.
   (b) We also don't need the last record. Just remember that the temperature the day of the Challenger accident was 31 degrees Fahrenheit. Remove the last record from the dataframe.
   (c) Extract the features in an numpy array called x and the labels in an numpy array called y. *Hint*: You can get the temperature by: `df['Temperature'].values`.
   (d) Perform logistic regression between the temperature (x) and the damage label (y). Do not bother doing a validation

because there are not a lot of data. Just use a very simple model so that you don't overfit.

(e) Plot the probability of damage as a function of the temperature.

(f) Decide whether or not to launch. The temperature the day of the Challenger accident was 31 degrees Fahrenheit. Would you go ahead with the launch or not? *Hint*: Start by calculating the probability of damage at 31 degrees Fahrenheit.

2. **Buckling Instability Classification (BIC1).** In this problem, you will work with the dataset BIC1.zip[3] of Professor Emma (Lejeune, 2021). I quote from the data repository:

> The Buckling Instability Classification (BIC) datasets contain the results of finite element simulations where a heterogeneous column is subject to a fixed level of applied displacement and is classified as either "Stable" or "Unstable." Each model input is a 16x1 vector where the entries of the vector dictate the Young's Modulus (E) of the corresponding portion of the physical column domain. Each input file has 16 columns one for each vector entry. For each 16x1 vector input, there is a single output that indicates if the column was stable or unstable at the fixed level of applied displacement. An output value of "0" indicates stable, and an output value of "1" indicates unstable. In BIC-1, we only allow two possible discrete values for E: $E = 1$ or $E = 4$. In BIC-2, we allow three possible discrete values for E: $E = 1$, $E = 4$, or $E = 7$. In BIC-3, we allow continuous values (to three digits of precision) of E in the range $E = 1$–8. BIC-1 consists of 65,536 simulation results. This exhausts the entire possible input domain. BIC-2 consists of 100,000 simulation results. This is less than 1% of the entire possible input domain. BIC-3 also consists of 100,000 simulation results. This is a tiny fraction of the entire possible input domain.

(a) Find, download and unzip the folder BIC1.zip from the data repository. There are four files in the folder: BIC1/train_input_data.txt, BIC1/test_input_data.txt, BIC1/train_output_data.txt, and BIC1/test_output_data.txt.

---

[3]Download from https://open.bu.edu/handle/2144/40085.

(b) Load the training input data into a numpy array called x_train and the training output data into a numpy array called y_train.

(c) Load the test input data into a numpy array called x_test and the test output data into a numpy array called y_test.

(d) Standardize the input data.

(e) Perform logistic regression between the input data (x) and the output data (y) using the training data. Use polynomial features of order 2. There are a lot of examples. If the problem is computationally too hard, you can either subsample the data or investigate how you can change the optimization algorithm to "sag" or "saga". *Hint*: Read the documentation of the LogisticRegression class of scikit-learn.

(f) Make predictions on the test data.

(g) Calculate the accuracy of the model on the test set.

(h) Plot the confusion matrix for the test set.

(i) Try to make a better model by creating new *features* from the original 16-dimensional input space. For example, you can experiment with the following:

- The percentage of with Young's Modulus $E = 1$ in the input vector.
- The number of switches from $E = 1$ to $E = 4$ in the input vector divided by 16.
- The number of switches from $E = 4$ to $E = 1$ in the input vector divided by 16.

3. **Buckling Instability Classification (BIC2).** Repeat the exercises above for the BIC2.zip dataset. Do not even bother with the polynomial features. It is unlikely that things will work. Try to come up with your own features. *Hint*: You can use radial basis functions with centers on some of the standardized inputs. You can do this using Support Vector Machines (SVM) as implemented in scikit-learn.[4]

---

[4]See the documentation of scikit-learn at https://scikit-learn.org/stable/modules/svm.html.

4. **Buckling Instability Classification (BIC3).** Repeat the exercises above for the BIC3.zip dataset. *Hint*: Try again SVM if you got it to work for BIC2. In addition, you can try building a Multi-Layer Perceptron (MLP).[5]

---

[5]See the scikit-learn documentation at https://scikit-learn.org/stable/modules /generated/sklearn.neural_network.MLPClassifier.html.

Bibliography

Aakash, B., Connors, J., and Shields, M. D. (2019). Stress-strain data for aluminum 6061-t651 from 9 lots at 6 temperatures under uniaxial and plane strain tension, *Data in Brief* **25**, p. 104085.

Bishop, C. M. (2006). *Pattern Recognition and Machine Learning* (Springer, New York).

Brodersen, K., Ong, C., Stephan, K., and Buhmann, J. (2010). The balanced accuracy and its posterior distribution, *Proceedings of the 20th International Conference on Pattern Recognition*.

Casey, A. D. (2020). Predicting energetic material properties and investigating the effect of pore morphology on shock sensitivity via machine learning, Purdue University Graduate School.

Jaynes, E. T. (2003). *Probability Theory: The Logic of Science* (Cambridge University Press, Cambridge).

Katsounaros, I., Dortsiou, M., Polatides, C., Preston, S., Kypraios, T., and Kyriacou, G. (2012). Reaction pathways in the electrochemical reduction of nitrate on tin, *Electrochimica Acta* **71**, pp. 270–276.

Lejeune, E. (2021). Geometric stability classification: Datasets, metamodels, and adversarial attacks, *Computer-Aided Design* **131**, p. 102948.

Pearl, J. (2019). The new science of cause and effect, YouTube video.

Raiffa, H. and Schlaifer, R. (1968). *Applied Statistical Decision Theory*, student edn. (M.I.T. Press, Cambridge, MA).

Silverman, B. W. (1985). Some aspects of the spline smoothing approach to nonparametric regression curve fitting, *Journal of the Royal Statistical Society: Series B (Methodological)* **47**, 1, pp. 1–21.

Smith, L. (1987). Los Alamos National Laboratory explosives orientation course: Sensitivity and sensitivity tests to impact, friction, spark and shock, Los Alamos National Lab, NM (USA).

Tufte, E. R. (2001). *The Visual Display of Quantitative Information*, 2nd edn. (Graphics Press, Cheshire, CT).

Wasserstein, R. L. and Lazar, N. A. (2016). The ASA statement on p-values: Context, process, and purpose, *The American Statistician* **70**, 2, pp. 129–133.

Index

credible intervals, 215, 223–224, 228–229, 237–238, 240, 322–323
cross-entropy loss, 329
cross-validation, 317–319, 325
CSV files, 43, 45–46, 52
cumulative distribution function (CDF), 171, 173–176, 178–180, 182–187, 216, 218–220, 222, 226–228

D

data cleaning, 51–52
DataFrame, 46–52
dataframe selection, 108–121
data loading, 37–52
data selection, 37–52
data visualization, 53–70
decision-making theory, 334
default parameters, 83–86
density, 65–66
describe, 46, 51
design matrix, 290, 293, 295, 297–301, 305, 307–308, 310, 321–322, 324
discrete random variables, 157–158
division, 7
docstrings, 78–79, 89, 92
domain knowledge, 1
dot product, 33
dropna, 51

E

entropy loss, 327
epistemic uncertainty, 124–125, 138–139
Euclidean norm, 33, 291–292
exhaustive, 152
expectation, 189–214
exponential distribution, 186–187
exponentiation, 7
extend, 29

F

false negatives, 337
false positives, 337

features, 327, 332, 337, 339
filtering, 2–3
finite differences, 59
fitting distributions, 229
floating point errors, 10
for loops, 101, 107
formatted string literals, 74
Fourier basis, 306–309, 316–317
f-strings, 74, 91
functions, 1, 8–13, 20

G

Gaussian distribution, 22, 215–221, 223–229
generalized linear model, 289, 302–304, 319
gradient descent, 59, 329

H

help documentation, 1
histograms, 53, 64–67

I

if-elif-else, 100
if-else, 100
if statements, 99, 106–107
iloc, 49, 55
immutable, 26
indentation, 99–100
independent, 264, 270–272, 275–276
independent random variables, 225, 241, 243, 271
indexing, 20–27
information flow, 154
insert, 28–29
integer division, 7
inverse CDF, 222

J

Jensen's inequality, 201
joint PDF, 241–245

K

keyword arguments, 84–86

www.ingramcontent.com/pod-product-compliance
Lightning Source LLC
Chambersburg PA
CBHW050537190326
41458CB00007B/1816

9789819822430